Lecture Notes in Control and Information Sciences

Edited by A. V. Balakrishnan and M. Thoma

Lecture Notes in Control and Information Sciences

Edited by A.V. Balakrishnan and M.Thoma

12

Ian Postlethwaite
Alistair G. J. MacFarlane

A Complex Variable Approach
to the Analysis of Linear
Multivariable Feedback Systems

Springer-Verlag
Berlin Heidelberg GmbH 1979

Authors
Dr. I. Postlethwaite,
Research Fellow, Trinity Hall, Cambridge, and
SRC Postdoctoral Research Fellow,
Engineering Department, University of Cambridge.

Professor A. G. J. MacFarlane,
Engineering Department, University of Cambridge,
Control and Management Systems Division,
Mill La'ie,
Cambridge CB2 1RX.

ISBN 978-3-540-09340-4   ISBN 978-3-540-35245-7 (eBook)
DOI 10.1007/978-3-540-35245-7

Originally published by Springer-Verlag Berlin Heidelberg New York in 1979.

2061/3020-543210

Contents

1. Introduction

The great success of the optimal control and optimal
filtering techniques developed for aerospace work during the
late 1950's and early 1960's naturally led to attempts to
apply these techniques to a wide range of earth-bound multivariable industrial processes.   In many situations this was
less than immediately successful, particularly in cases where
the available plant models were not sufficiently accurate or
where the performance indices required to stipulate the controlled
plant behaviour were much less obvious in form than in the aerospace context. Moreover, the controller which results from a
direct application of optimal control and optimal filtering
synthesis techniques is in general a complicated one; in fact,
if it incorporates a full Kalman-Bucy filter it has a
dynamical complexity equal to that of the plant which it is
controlling, since the filter essentially consists of a plant
model with feedback around it. In contrast, what was needed
for many multivariable process control problems was a relatively
simple controller which would both stabilize, about an operating
point, a plant for which only a very approximate model might be
available, and also mitigate the effect of low-frequency
disturbances by incorporating integral action.   To industrial
engineers brought up on frequency-response ideas the sophisticated
optimal control methods seemed difficult to use; these engineers
essentially relied on a mixture of physical insight and straightforward techniques, such as the use of derivative and integral
action, to solve their problems.   It became obvious that a
huge gap in techniques existed between the classical single-
loop frequency-response methods, based on the work of Nyquist [1],

Bode [2] and Evans [3] which were still in use for many
industrial applications, and the elegant and powerful multivariable time-response methods developed for aerospace
applications.

For these reasons an interest in frequency-response
methods slowly began to revive during the mid-1960's. An
important first step towards closing the yawning gap between
an optimal control approach and the classical frequency-response
approach was taken by Kalman [4] , who studied the frequency-
domain characterization of optimality. A systematic attack
on the whole problem of developing a frequency-response analysis
and design theory for multivariable systems was begun in a
pioneering paper by Rosenbrock [5] which ushered in a decade
of increasing interest in a rejuvenated frequency-response
approach. Prior to this new point-of-departure some fairly
straightforward attacks had been made on the multivariable
control problem. Boksenbom and Hood [6] put forward the
idea of a non-interacting controller. Their procedure consisted
simply of choosing a cascaded compensator such that the overall
transfer function matrix of the compensated system had a
diagonal form. If such a compensator could be found then the
controller design could be finished off using standard single-
loop design techniques. The required compensating matrix which
usually results from such a procedure is necessarily a complicated
one, and the most succinct objection to this approach is simply
that it is not essential to go to such drastic lengths merely
to reduce interaction. A natural further step in this initial
approach to multivariable control was to see what could be
achieved by way of standard matrix calculations using rational
matrices; papers studying the problem in this way were produced

by Golomb and Usdin [7] , Raymond [8] ; Kavanagh [9] , [10],[11] , and Freeman [12] , [13] . Rosenbrock [14],[15], however, opened up a completely new line of development by seeking to reduce a multi-variable problem to one amenable to classical techniques in a more sophisticated way. In his Inverse Nyquist Array Method [14] , [15] the aim was to <u>reduce</u> interaction to an amount which would then enable single-loop techniques to be employed, rather than to eliminate interaction completely. The Rosenbrock approach was based upon a careful use of a specific criterion of partial interaction - the diagonal dominance concept. The success of this method led other investigators to develop ways of seeking to reduce a multivariable control problem to a succession of single-loop problems, as in the Sequential Return Difference approach of Mayne [16].

In the non-interacting, or partially non-interacting, approach to multivariable control the motivation was the eventual deployment of classical single-loop frequency-response techniques during the final stages of a design study. An alternative approach, however, is to investigate the transfer-function matrix representation as a single object in its own right and to ask : how can the key basic concepts of the classical single-loop frequency-response approach be suitably extended? What are the relevant generalizations to the multivariable case of the specific concepts of pole, zero, Nyquist diagram and root locus diagram? It is to questions of this sort that the work presented here is addressed, and it is shown that complex-variable ideas have an important role to play in the study of multivariable feedback systems. An early attempt to extend Nyquist diagram ideas to the multivariable problem was made by Bohn [17] , [18] . A generalization of the Nyquist stability criterion was put forward by MacFarlane [19] and, following that heuristic treatment, complex-variable based proofs were supplied by Barman and Katzenelson [20] and MacFarlane and Postlethwaite [21] . This generalization of the Nyquist stability criterion to the multivariable situation was soon followed by complementary generalizations of the root locus technique [21], [22], [23], [24].

The aim of the work presented in this text is to extend the concepts underlying the techniques of Nyquist, Bode and Evans to multivariable systems. In the two classical approaches to linear feedback system design the Nyquist-Bode approach studies gain as a function of frequency and the Evans' approach studies frequency as a function of gain. In Chapter 3 it is shown how the ideas of studying complex gain as a function of complex frequency and complex frequency as a function of complex gain can be extended to the multivariable case by associating with transfer function matrices (having the same number of rows and columns) a pair of analytic functions : a characteristic gain function and a characteristic frequency function. These are algebraic functions [25] and each is defined on an appropriate Riemann surface [26]. Chapter 2 deals with a number of essential preliminaries such as a description of the type of multivariable feedback system being considered; with basic definitions of stability and related theorems; and with a fundamental relationship between open- and closed-loop behaviour based on the return-difference operator. Chapter 3 also contains a comprehensive discussion of the background to the generalized Nyquist stability criterion for multivariable feedback systems which is presented in Chapter 4. The proof of this criterion is based on the Principle of the Argument applied to an algebraic function defined on an appropriate Riemann surface. In Chapter 5 a generalization of the inverse Nyquist stability criterion to the multivariable case is developed which is complement-ary to the exposition of the generalized Nyquist criterion given in the previous chapter. Using the material developed in Chapter 3, the Evans' root locus approach is extended to multivariable systems

in Chapter 6; this uses well established results in algebraic
function theory. It is also shown how an algebraic-function
based approach can be used to find the asymptotic behaviour of
the closed-loop poles of a multivariable time-invariant optimal
linear regulator as the weight on the input terms of a quadratic
performance index approaches zero.

As the work presented progresses it becomes evident that
the gain variable used can be considered as a parameter of the
system, and consequently that the techniques developed are not
only applicable to gain and frequency but to any parameter and
frequency. In Chapter 7 the effect of parameter variations
on a multivariable feedback system is considered by the introduction of the concepts of 'parametric' root loci and 'parametric'
Nyquist loci. This chapter concludes with a few tentative
proposals and suggestions for future research.

Information of secondary importance which would unnecessar-
ily break the flow of the text has been placed in appendices.
References are listed at the end of each chapter in which they
are cited, and also at the end of the text where a bibliography
is provided.

References

[1]   H.Nyquist, "Regeneration theory", Bell Syst. Tech.J.,
       11, 126-147, 1932.

[2]   H.W.Bode, "Network analysis and feedback amplifier design",
       Van Nostrand, Princeton, N.J., 1945.

[3]   W.R.Evans, "Graphical analysis of control systems", Trans.
       AIEE, 67, 547-551, 1948.

[4]   R.E.Kalman, "When is a linear control system optimal?",
       Trans. ASME J.Basic Eng., Series D., 86, 51-60, 1964.

[5]   H.H.Rosenbrock, "On the design of linear multivariable
       control systems", Proc. Third IFAC Congress London, 1,
       1-16, 1966.

[6]   A.S.Boksenbom and R.Hood, "General algebraic method
         applied to control analysis of complex engine types",
         National Advisory Committee for Aeronautics, Report
         NCA-TR-980, Washington D.C., 1949.

[7]   M.Golomb and E.Usdin, "A theory of multidimensional
         servo systems", J.Franklin Inst., 253(1), 28-57, 1952.

[8]   F.H.Raymond, "Introduction a l'étude des asservissements
         multiples simultanes", Bull. Soc. Fran. des Mecaniciens,
         7, 18-25, 1953.

[9]   R.J.Kavanagh, "Noninteraction in linear multivariable
         systems", Trans. AIEE, 76, 95-100, 1957.

[10]   R.J.Kavanagh, "The application of matrix methods to
         multivariable control systems", J.Franklin Inst., 262,
         349-367, 1957.

[11]   R.J.Kavanagh, "Multivariable control system synthesis",
         Trans. AIEE, Part 2, 77, 425-429, 1958.

[12]   H.Freeman, "A synthesis method for multipole control
         systems", Trans. AIEE, 76, 28-31, 1957.

[13]   H.Freeman, "Stability and physical realizability con-
         siderations in the synthesis of multipole control systems",
         Trans.AIEE, Part 2, 77, 1-15, 1958.

[14]   H.H.Rosenbrock, "Design of multivariable control systems
         using the inverse Nyquist array, Proc.IEE, 116, 1929-1936,
         1969.

[15]   H.H.Rosenbrock, "Computer-aided control system design",
         Academic Press, London, 1974.

[16]   D.Q.Mayne, "The design of linear multivariable systems",
         Automatica, 9, 201-207, 1973.

[17]   E.V.Bohn, "Design and synthesis methods for a class of
         multivariable feedback control systems based on single
         variable methods", Trans.AIEE, 81, Part 2, 109-115, 1962.

[18]   E.V.Bohn and T.Kasvand, "Use of matrix transformations and
         system eigenvalues in the design of linear multivariable
         control systems", Proc.IEE, 110, 989-997, 1963.

[19]   A.G.J.MacFarlane, "Return-difference and return-ratio
         matrices and their use in the analysis and design of
         multivariable feedback control systems", Proc.IEE, 117,
         2037-2049, 1970.

[20]   J.F.Barman and J.Katzenelson, "A generalized Nyquist-
         type stability criterion for multivariable feedback
         systems", Int.J.Control, 20, 593-622, 1974.

[21]   A.G.J.MacFarlane and I. Postlethwaite, "The generalized
        Nyquist stability criterion and multivariable root loci",
        Int. J. Control, 25, 81-127, 1977.

[22]   B.Kouvaritakis and U.Shaked, "Asymptotic behaviour of
        root loci of linear multivariable systems", Int. J.
        Control, 23, 297-340, 1976.

[23]   I. Postlethwaite, " The asymptotic behaviour, the
        angles of departure, and the angles of approach of
        the characteristic frequency loci", Int. J. Control,
        25, 677-695, 1977.

[24]   A.G.J.MacFarlane, B.Kouvaritakis and J.M.Edmunds,
        "Complex variable methods for multivariable feedback
        systems analysis and design", Alternatives for Linear
        Multivariable Control, National Engineering Consortium,
        Chicago, 189-228, 1977.

[25]   G.A.Bliss, "Algebraic functions", Dover, New York,
        1966 (Reprint of 1933 original).

[26]   G. Springer, "Introduction to Riemann surfaces",
        Addison-Wesley, Reading, Mass., 1957.

2. Preliminaries

This text considers the generalization of the classical techniques of Nyquist and Evans to a linear time-invariant dynamical feedback system which consists of several multi-input, multi-output subsystems connected in series.   In this chapter a description of the multivariable feedback system under consideration is given.   The chapter also includes basic definitions of stability, some associated theorems, and a fundamental relationship between open- and closed-loop behaviour based on the return-difference operator.

2.1 System description

The basic description of a linear time-invariant dynamical system is taken to be the state-space model

$$\dot{x}(t) = Ax(t) + Bu(t)$$
$$y(t) = Cx(t) + Du(t) \qquad (2.1.1)$$

where $x(t)$ is the state vector, $y(t)$ the output vector, $u(t)$ the input vector; $\dot{x}(t)$ denotes the derivative of $x(t)$ with respect to time; $A,B,C,$ and D are constant real matrices.   For convenience the model will be denoted by $S(A,B,C,D)$ or S when the meaning is obvious, and represented diagramatically as shown in figure 1.

In general $S(A,B,C,D)$ will be considered as being the state-space representation of several subsystems

$$S_i(A_i,B_i,C_i,D_i): \quad \dot{x}_i(t) = A_i x_i(t) + B_i u_i(t)$$
$$i=1,2,\ldots\ldots,h \qquad y_i(t) = C_i x_i(t) + D_i u_i(t) \qquad (2.1.2)$$

connected in series, as illustrated in figure 2.   If, for example, S consists of two subsystems S_1 and S_2 then the

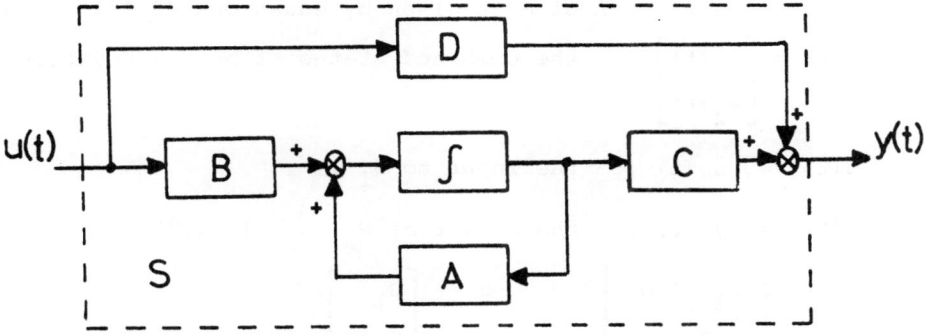

Figure 1. State-space model

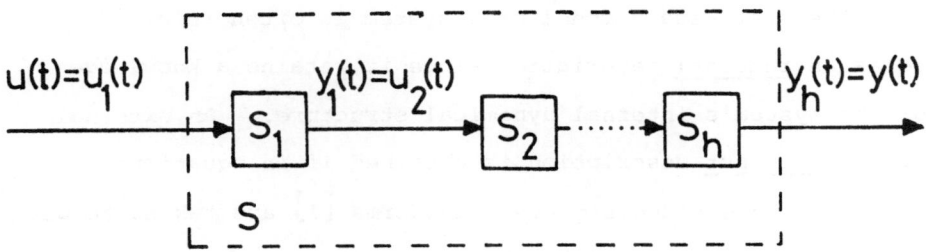

Figure 2. Series connection

of subsystems

state-space description of S is given by equations (2.1.1) with $x(t) = \begin{bmatrix} x_1(t) \\ x_2(t) \end{bmatrix}$, the combined states of both subsystems,

$u(t) = u_1(t)$, the input to S_1,

$y(t) = y_2(t)$, the output of S_2, $\qquad$ (2.1.3)

$A = \begin{bmatrix} A_1 & O \\ B_2 C_1 & A_2 \end{bmatrix}$, $\qquad B = \begin{bmatrix} B_1 \\ B_2 D_1 \end{bmatrix}$,

$C = \begin{bmatrix} D_2 C_1 & C_2 \end{bmatrix}$, and $D = D_2 D_1$.

The state-space model for an interconnection of several subsystems can be derived from the above formula by successive application.

The state-space model of a system is often referred to as an _internal_ description since it retains a knowledge of the system's internal dynamical structure. An _external_ or _input-output_ description is obtained if in equations (2.1.1) single-sided Laplace transforms [1] are taken, to give

$$s\hat{x}(s) - x(o) = A\hat{x}(s) + B\hat{u}(s)$$
$$\hat{y}(s) = C\hat{x}(s) + D\hat{u}(s)$$
$\qquad$ (2.1.4)

where $\hat{x}(s)$ denotes the Laplace transform of $x(t)$. If the initial conditions at time $t=0$ are all zero so that $x(o)=0$, then the input and output transform vectors are related by

$$\hat{y}(s) = G(s) \, \hat{u}(s) \qquad (2.1.5)$$

where

$$G(s) = C(sI_n - A)^{-1}B + D \qquad (2.1.6)$$

I_n is a unit matrix of order n and $(\;)^{-1}$ denotes the inverse of a matrix. $G(s)$ is a matrix-valued rational function of the complex variable s, and is called the <u>transfer</u>

<u>function matrix</u> for the set of input-output transforms, or

the <u>open-loop gain matrix</u>.   The transfer function matrix

G(s) can be regarded as describing a system's response to

an exponential input with exponent s [2], and therefore

the complex variable s can be considered a complex

frequency variable as in the single-input, single-output case.

When S(A,B,C,D) consists of h subsystems

$\{S_i(A_i,B_i,C_i,D_i):   i=1,2,....,h\}$, as shown in figure 2,

each subsystem has a transfer function matrix

$$G_i(s) = C_i(sI_{n_i}-A_i)^{-1}B_i + D_i \qquad (2.1.7)$$

and the input-output transform vectors of S are related by

$$\hat{y}(s)=G_h(s)G_{h-1}(s)....G_1(s) \, \hat{u}(s) \qquad (2.1.8)$$

with the obvious relationship for the open-loop gain matrix

$$G(s) = G_h(s)G_{h-1}(s)....G_1(s) \qquad (2.1.9)$$

For the purpose of connecting outputs back to inputs to

form a feedback loop G(s) is assumed to be a square matrix

of order m.

2.2   Feedback configuration

The general feedback configuration that will be considered

is shown in figure 3.   The output of the feedback system

is shown as that of the hth subsystem but in practice it

may be the output from an earlier subsystem in which case

the later subsystems can be thought of as being feedback

compensators.   The parameter k is a real gain control

variable common to all the loops.   The system's input and

12

output are related to the reference input $r(t)$ by the equations

$$e(t) = r(t) - y(t)$$

$$u(t) = ke(t)$$

(2.2.1)

and combining these with equations (2.1.1) the following closed-loop state-space equations are obtained:

$$\dot{x}(t) = A_c x(t) + B_c r(t)$$

$$y(t) = C_c x(t) + D_c r(t)$$

(2.2.2)

where

$$A_c = A - B(k^{-1}I_m + D)^{-1}C$$

$$B_c = kB - kB(k^{-1}I_m + D)^{-1}D$$

$$C_c = (I_m + kD)^{-1}C$$

$$D_c = (k^{-1}I_m + D)^{-1}D$$

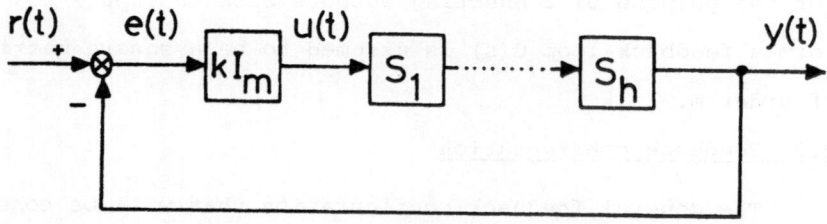

Figure 3. Feedback configuration

2.3   Stability

Stability is the most important single requirement
of a feedback system and for general time-dependent
nonlinear systems it poses very complex problems.   The
stability problem for linear time-invariant dynamical
systems, however, is much simpler than in the general case.
This is because:

(i)   all stability properties are constant with respect
to time, and

(ii)   all stability properties are global, since any solution
for the state of the system is proportional to the state
at time zero;   see equation (2.3.2).

There are many definitions of stability in the literature
and broadly speaking these can be divided into two classes.
The first class of definitions concerns stability of free
systems i.e. those in which there is no input;   the second
class of definitions concerns the behaviour of forced
systems i.e. those in which there is a given input.   Both
types of stability are discussed below, the definitions and
associated theorems following very closely those given by
Willems [3] .

2.3-1   Free systems

Let us consider the closed-loop dynamical system of
figure 3 described by equations (2.2.2) with r(t)=0 and
C_c=I.   Then the stability problem reduces to that of
considering the free system

$$\dot{x}(t) = A_c x(t) \tag{2.3.1}$$

The equilibrium state for equation (2.3.1) is clearly the

origin (assuming A_c is non-singular), and therefore a
solution of (2.3.1) which passes through the origin at some
time remains there for all subsequent times; this solution
is called the null solution. The stability of the origin
equilibrium state is characterized using the following
definitions.

<u>Definition 1</u>. The origin of the free system (2.3.1) is
called stable if when the system is perturbed from the
origin all subsequent motions remain in a correspondingly
small neighbourhood of the origin.

<u>Definition 2</u>. The origin of the free system (2.3.1) is
called asymptotically stable if when the system is perturbed
slightly from the origin all subsequent motions return to
the origin.

<u>Definition 3</u>. The origin of the free system (2.3.1) is
called asymptotically stable in the large, or globally
asymptotically stable, if it is stable, and if every
motion converges to the origin as $t \to \infty$.

The general solution of equation (2.3.1) is [3]

$$x(t ; x(t_o), t_o) = \exp \left[A_c(t-t_o) \right] . x(t_o) \qquad (2.3.2)$$

which shows clearly that if the free system is asymptotically
stable it is also asymptotically stable in the large. If J
is the Jordan canonical form [4] of A_c such that

$$A_c = TJT^{-1} \qquad (2.3.3)$$

with

$$J = \begin{bmatrix} J_1 & & & \\ & J_2 & & \\ & & \ddots & \\ & & & J_k \end{bmatrix}$$

where each Jordan block J_i has the form

$$J_i = \begin{bmatrix} \lambda_i & 1 & & & \\ & \lambda_i & \ddots & & \\ & & \ddots & 1 & \\ & & & \lambda_i & 1 \\ & & & & \lambda_i \end{bmatrix}$$

and λ_i is an eigenvalue of A_c, then it can be shown $[3]$, that

$$\exp\left[A_c(t-t_o)\right] = T \exp\left[J(t-t_o)\right].T^{-1} \qquad (2.3.4)$$

with

$$\exp\left[J(t-t_o)\right] = \begin{bmatrix} \exp\left[J_1(t-t_o)\right] & & & \\ & \exp\left[J_2(t-t_o)\right] & & \\ & & \ddots & \\ & & & \exp\left[J_k(t-t_o)\right] \end{bmatrix}$$

and

$$\exp\left[J_i(t-t_o)\right] = \begin{bmatrix} 1 & \underline{t} & \underline{t}^2/2! & \cdots & \underline{t}^{r-1}/(r-1)! \\ 0 & 1 & \underline{t} & & \underline{t}^{r-2}/(r-2)! \\ \vdots & \vdots & \vdots & & \vdots \\ 0 & 0 & 0 & & 1 \end{bmatrix} \exp\left[\lambda_i(t-t_o)\right]$$

where $\underline{t} = (t-t_o)$

and r is the order of the Jordan block J_i. The general

solution of the free system can therefore be expressed as

$$x(t; x(t_o),t_o) = T \exp\left[J(t-t_o)\right].T^{-1}x(t_o) \qquad (2.3.5)$$

and from this the following theorems can be derived; see

$[3]$ for proofs.

Theorem 1. The null solution of system (2.3.1) is asymptotically

stable if and only if all eigenvalues of the matrix A_c have

negative real parts.

Theorem 2. The null solution of system (2.3.1) is stable

if and only if the matrix A_c has no eigenvalues with positive

real parts, and if the eigenvalues with zero real parts

correspond to Jordan blocks of order 1.

2.3-2   Forced systems

Let us consider the closed-loop dynamical system (2.2.2) which has the general solution [3] ,

$$x(t; x(t_o),t_o) = \exp (A_c t).x(t_o)+\int_o^t \exp[A_c(t-\tau)].B_c r (\tau) d\tau$$

$$(2.3.6)$$

To study the stability properties of this system we need to introduce the concept of input-output stability.

__Definition 4__.   A dynamical system is called input-output stable if for any bounded input a bounded output results regardless of the initial state.

By theorem 1, asymptotic stability of the unforced system (2.3.1) implies that all the eigenvalues of A_c have negative real parts, in which case there exist positive numbers P and a such that

$$\|\exp(A_c t)\| \leqslant P \exp (-at) \qquad \forall t \geqslant 0 \qquad (2.3.7)$$

where $\|.\|$ denotes the Euclidean norm of a matrix or vector [3].   From equations (2.2.2), (2.3.6) and (2.3.7) we then have

$$\| y(t)\| \leqslant \quad \| C_c x(t)\| \quad + \quad \| D_c r(t)\|$$

$$\leqslant \quad \| D_c r(t)\| \quad + \quad \| C_c \exp(A_c t).x(t_o)\|$$

$$+ c\int_o^t \|\exp[A_c(t-\tau)]\| \|B_c r(t)\| d\tau$$

$$\leqslant \quad dM+cP\|x(t_o)\| + cbMP/a$$

where $b=\|B_c\|$, $c=\|C_c\|$, $d=\|D_c\|$, and $\|r(t)\|\leqslant M \; \forall \; t \geqslant 0$.

This result is summarized in the following theorem.

__Theorem 3__.   If the null solution of the unforced system (2.3.1) is asymptotically stable, then the forced system (2.2.2) is input-output stable.

Note that input -output stability implies asymptotic stability

of the equilibrium state at the origin only if the system
(2.2.2) is state controllable and state observable; or if
all unobservable and/or uncontrollable modes have negative
real parts.   In the remainder of this book system stability
is understood as meaning input-output stability coupled with
asymptotic stability of the equilibrium state at the origin.

Theorem 3 is important because it tells us that the
stability of a linear time-invariant system can be determined
solely from a knowledge of the eigenvalues of the system "A"
matrix.   The stability conscious eigenvalues corresponding
to the closed-loop dynamical system (2.2.2) are values of λ
which satisfy the equation

$$\det\left[\lambda I_n - A_c\right] = 0 \qquad (2.3.8)$$

The left-hand side of equation (2.3.8) is called the
closed-loop characteristic polynomial, abbreviated as CLCP(λ)
so that

$$CLCP(\lambda) \triangleq \det\left[\lambda I_n - A_c\right] \qquad (2.3.9)$$

Similarly for the open-loop system S(A,B,C,D) an open-loop
characteristic polynomial, OLCP(λ), is defined as

$$
\begin{aligned}
OLCP(\lambda) &\triangleq \det\left[\lambda I_n - A\right] \\
&= \det\left[\lambda I_{n_1} - A_1\right] \det\left[\lambda I_{n_2} - A_2\right] \dots \\
&\quad \dots \det\left[\lambda I_{n_h} - A_h\right] \qquad (2.3.10)
\end{aligned}
$$

In the next section it is shown how the open- and
closed-loop characteristic polynomials are related via the
return-difference operator [5].

2.4.   Relationship between open- and closed-loop characteristic
polynomials for the general feedback configuration

Let us suppose that all the feedback loops of the general

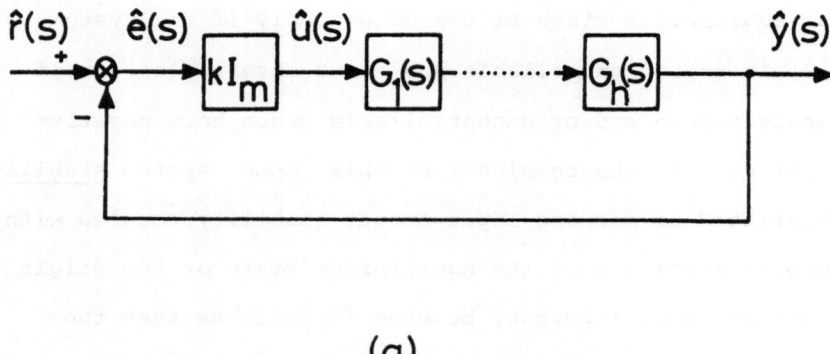

(a)

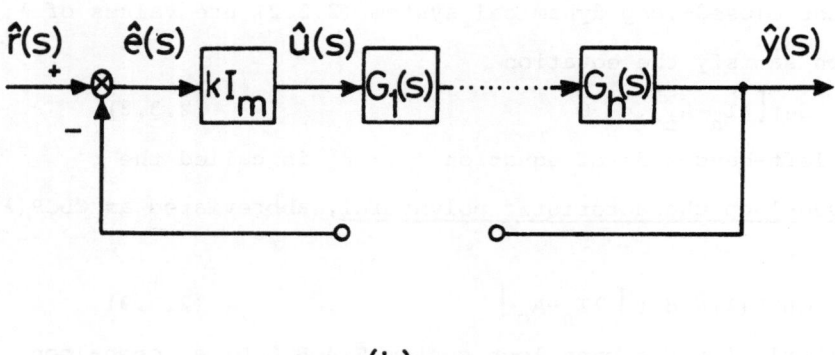

(b)

Figure 4. Feedback configuration
(a) closed-loop
(b) open-loop

closed-loop configuration are broken and that the subsystems
are represented by their transfer function matrices; see
figure 4.   The corresponding return-difference matrix
[5] for this break point is

$$F(s) \triangleq I_m + L(s) \qquad\qquad (2.4.1)$$

where

$$L(s) = kG_h(s)G_{h-1}(s)\ldots.G_1(s)$$
$$= kG(s) \qquad\qquad (2.4.2)$$

is called the system return-ratio matrix [5] .   A return-
difference operator generates the difference between injected
and returned signal transforms from the injected signal
transform.   It plays a major role in feedback theory since
the essence of forging a feedback link is making two sets
of signals identically equal, thus making the difference
between them identically zero.   Both F(s) and L(s) are
matrix-valued rational functions of a complex variable and
the key concepts in this text revolve around the properties
of such matrices.   The importance of the return-difference
matrix is emphasized in the relationship between open- and
closed-loop characteristic polynomials which is now derived
for the general feedback configuration.

If we take determinants of equation (2.4.1) and represent
G(s) by its state-space model, we obtain

$$\det F(s) = \det\left[I_m + kC(sI_n - A)^{-1}B + kD\right] \qquad (2.4.3)$$

which using Schur's formula [6] for the evaluation of
partitioned determinants can be rewritten as

$$\det F(s) = \det\left[\begin{array}{c|c} sI_n - A & B \\ \hline -kC & I_m + kD \end{array}\right] \div \det\left[sI_n - A\right] \qquad (2.4.4)$$

which is equivalent to

$$\det F(s) = \det \begin{bmatrix} I_n & \vdots & -B(I_m+kD)^{-1} \\ 0 & \vdots & I_m \end{bmatrix} \det \begin{bmatrix} sI_n-A & \vdots & B \\ -kC & \vdots & I_m+kD \end{bmatrix} \div \det [sI_n-A]$$

$$= \det \begin{bmatrix} sI_n-A+B(k^{-1}I_m+D)^{-1}C & \vdots & 0 \\ -kC & \vdots & I_m+kD \end{bmatrix} \div \det [sI_n-A]$$

$$= \frac{\det [sI_n-A+B(k^{-1}I_m+D)^{-1}C] \det [I_m+kD]}{\det [sI_n-A]} \qquad (2.4.5)$$

Now from equations (2.2.2) we have

$$A_c = A-B(k^{-1}I_m+D)^{-1}C$$

and it is obvious from equation (2.4.3) that

$$\det F(\infty) = \det [I_m+kD]$$

and therefore under the assumption that $\det F(\infty) \neq 0$ we have

from equation (2.4.5) the following relationship

$$\boxed{\frac{\det F(s)}{\det F(\infty)} = \frac{\det [sI_n-A_c]}{\det [sI_n-A]} = \frac{\det [sI_n-A_c]}{\det [sI_{n_h}-A_h]\dots\det [sI_{n_1}-A_1]} \triangleq \frac{CLCP(s)}{OLCP(s)}}$$

$$(2.4.6)$$

The zeros of the open- and closed-loop characteristic
polynomials, OLCP(s) and CLCP(s), are known as the open-
and closed-loop poles or characteristic frequencies respectively.

Relationship (2.4.6) shows how the matrix-valued
rational transfer functions F(s) and G(s) are intimately
related to the stability of a dynamical feedback system.
The study of such matrices and their eigenvalues opens the way
to suitable extensions of the classical techniques of Nyquist
[7] and Evans [8;9] ; the results of such a study are given
in Chapter 3.

References

[1]   R. Bracewell, "The Fourier Transform and Its Applications",
      McGraw-Hill, New York, 1965.

[2] A.G.J. MacFarlane and N. Karcanias, "Poles and zeros of linear multivariable systems: a survey of the algebraic, geometric and complex variable theory", Int. J. Control, 24, 33-74, 1976.

[3] J.L. Willems, "Stability Theory of Dynamical Systems", Nelson, London, 1970.

[4] P.M. Cohn, "Algebra", Vol. 1, Wiley, London, 1974.

[5] A.G.J. MacFarlane, "Return-difference and return-ratio matrices and their use in analysis and design of multivariable feedback control systems", Proc. IEE, 117, 2037-2049, 1970.

[6] F.R. Gantmacher, "Theory of Matrices", Vol. 1, Chelsea, New York, 1959.

[7] H. Nyquist, "The Regeneration Theory", Bell System Tech. J., 11, 126-147, 1932.

[8] W.R. Evans, "Graphical Analysis of Control Systems", Trans. AIEE, 67, 547-551, 1948.

[9] W.R. Evans, "Control System Synthesis by Root Locus Method", Trans. AIEE, 69, 1-4, 1950.

3.   Characteristic gain functions and
        characteristic frequency functions

In the analysis and design of linear single-loop
feedback systems the two classical approaches use complex
functions to study open-loop gain as a function of imposed
frequency (the Nyquist-Bode approach), and to study closed-
loop frequency as a function of imposed gain (the Evans
root locus approach).   The primary purpose of this chapter
is to show how these techniques can be extended to the multivariable case by associating with appropriate matrix-
valued rational functions of a complex variable characteristic
gain functions and characteristic frequency functions.

3.1   Duality between open-loop gain and closed-loop frequency

For the general feedback configuration of figure 4 we
have from section 2.4 the fundamental relationship

$$\frac{\det F(s)}{\det F(\infty)} = \frac{\det[sI_n - A_c]}{\det[sI_n - A]} \tag{3.1.1}$$

where the return-difference matrix $F(s)$ is given as

$$F(s) = I_m + kG(s) \tag{3.1.2}$$

If we substitute for $F(s)$ in equation (3.1.1) we obtain

$$\frac{\det[sI_n - A_c]}{\det[sI_n - A]} = \frac{\det[I_m + kG(s)]}{\det[I_m + kD]}$$

$$= \frac{\det[k^{-1}I_m + G(s)]}{\det[k^{-1}I_m + D]} \tag{3.1.3}$$

and substituting for the gain variable k using the expression

$$g = -\frac{1}{k} \tag{3.1.4}$$

where g is allowed to be complex i.e. $g \; \varepsilon \; \mathbb{C}$ (the complex
plane), we have

$$\frac{\det\left[sI_n - A_c\right]}{\det\left[sI_n - A\right]} = \frac{\det\left[gI_m - G(s)\right]}{\det\left[gI_m - D\right]} \tag{3.1.5}$$

The closed-loop system matrix A_c is given in equations (2.2.2) as

$$A_c = A - B(k^{-1}I_m + D)^{-1}C \tag{3.1.6}$$

and substituting for k from equation (3.1.4) we have

$$A_c = A + B(gI_m - D)^{-1}C$$
$$\triangleq S(g) \tag{3.1.7}$$

The expression (3.1.5) can therefore be rewritten as

$$\boxed{\frac{\det\left[sI_n - S(g)\right]}{\det\left[sI_n - A\right]} = \frac{\det\left[gI_m - G(s)\right]}{\det\left[gI_m - D\right]}}$$

or

$$\boxed{\frac{\det\left[sI_n - S(g)\right]}{\det\left[sI_n - S(\infty)\right]} = \frac{\det\left[gI_m - G(s)\right]}{\det\left[gI_m - G(\infty)\right]}} \tag{3.1.8}$$

The form of this relationship shows a striking 'duality' between the complex frequency variable s and the complex gain variable g via their 'parent' matrices $S(g)$ and $G(s)$ respectively. This duality between the roles of frequency and gain forms the basis on which the classical complex variable methods are generalized to the multivariable case. $S(g)$ is called the <u>closed-loop frequency matrix</u>; its eigenvalues are the <u>closed-loop characteristic frequencies</u> and are clearly dependent on the gain variable g. The eigenvalues of the <u>open-loop gain matrix</u> $G(s)$ are called <u>open-loop characteristic gains</u> and are clearly dependent on the frequency variable s. The similarity between $G(s)$ and $S(g)$ is stressed if one examines their state-space structures:

$$G(s) = C(sI_n - A)^{-1}B + D \tag{3.1.9}$$

$$S(g) = B(gI_m - D)^{-1}C + A \tag{3.1.10}$$

24

In figure 5 the feedback configuration of figure 4a is redrawn with zero reference input, the state-space representation for G(s), and the substitution (3.1.4) for k in order to illustrate explicity the duality between the closed-loop characteristic frequency variable s and the open-loop characteristic gain variable g.

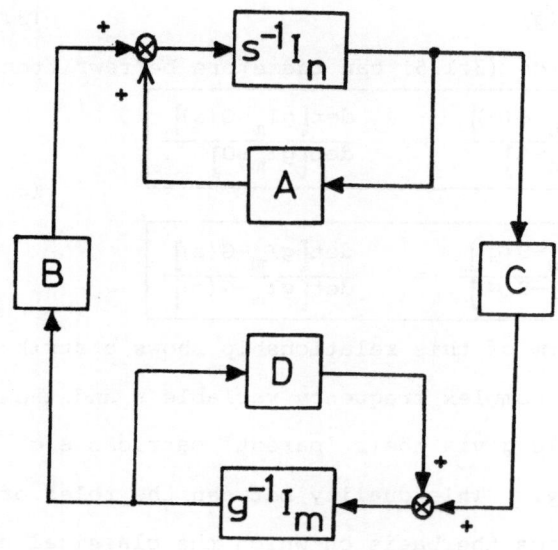

Figure 5. Feedback configuration illustrating the duality between s and g

The importance of relationship (3.1.8) is that it shows, for values of $s \notin \sigma$ (A) and values of $g \notin \sigma$(D) (this condition is equivalent to det $F(\infty) \neq 0$ which has already been assumed), where σ(A) denotes the spectrum of A , that

$$\det\left[sI_n - S(g)\right] = 0 \iff \det\left[gI_m - G(s)\right] = 0 \qquad (3.1.11)$$

This tells us that a knowledge of the open-loop characteristic gain as a function of frequency is equivalent to a knowledge of closed-loop characteristic frequency as a function of gain. The inference from this is that it ought to be possible to determine the stability of a feedback system from a knowledge of the characteristic gain spectrum of G(s). Note that from equation (3.1.8) we have that

$$CLCP(s) = \frac{\det\left[gI_m - G(s)\right]}{\det\left[gI_m - D\right]} \cdot OLCP(s) \qquad (3.1.12)$$

and such an expression makes it intuitively obvious that there should be a generalization of Nyquist's stability theorem to loci of the characteristic gains of G(s) as a function of frequency.

3.2 Algebraic functions: characteristic gain functions and characteristic frequency functions.

The characteristic equations for G(s) and S(g) i.e.

$$\Delta(g,s) \triangleq \det\left[gI_m - G(s)\right] = 0 \qquad (3.2.1)$$

and

$$\nabla(s,g) \triangleq \det\left[sI_n - S(g)\right] = 0 \qquad (3.2.2)$$

are algebraic equations relating the complex variables s and g. Each equation can be considered as a polynomial in g or s with coefficients which are rational functions in s or g respectively, and if irreducible over the field of rational functions each equation defines a pair of algebraic functions [1; appendix 1]:

(i) a <u>characteristic gain function g(s)</u> which gives open-loop characteristic gain as a function of frequency, and

(ii) a <u>characteristic frequency function s(g)</u> which gives closed-loop characteristic frequency as a function of gain.

In general equations (3.2.1) and (3.2.2) will not be irreducible and each equation will define a set of characteristic gain and characteristic frequency functions. For simplicity of exposition and because this is in any case the usual situation for G(s) and S(g) arising from practical situations, it will normally be assumed that equations (3.2.1) and (3.2.2) are irreducible over the field of rational functions.

Although both equation (3.2.1) and equation (3.2.2) define the same functions g(s) and s(g), equation (3.2.2) will in general contain more information about the system. It is possible under certain circumstances that $\nabla(s,g)$ will contain factors of s independent of g which are not present in $\Delta(g,s)$. These factors occur in the following situations:-

(1)   When the A-matrix of the open-loop system S(A,B,C,D) has eigenvalues which correspond to modes of the system which are unobservable and/or uncontrollable from the point of view of considering the input as that of the first sub-system and the output as that of the hth sybsystem. Note that if output measurements for earlier subsystems are available then in practice some of the unobservable modes of S(A,B,C,D) may in fact be observable.

(2)   When the poles and zeros of the open-loop gain matrix G(s) are different from the poles and zeros of the characteristic gain function g(s);   see section 3.3-3.

These two conditions, under which equations (3.2.1) and (3.2.2) differ, clearly present problems to the development

of a Nyquist-like stability criterion in terms·of loci of
the characteristic gains of G(s).   However, by relating the
poles and zeros of g(s) to the poles and zeros of G(s), and
by careful consideration of the unobservable and uncontrollable
modes these problems can be overcome; a generalized Nyquist
stability criterion is developed in chapter 4.

In the next section a detailed study of the characteristic
gain function is given which results in a generalization
of the root locus diagram.   In section 3.4 a similar study
of the characteristic frequency function results in a
generalized Nyquist diagram.

3.3   Characteristic gain functions

The natural way to define the characteristic gain function $g(s)$
is   via the characteristic equation

$$\Delta(g,s) \overset{\Delta}{=} \det\left[gI_m - G(s)\right] = 0 \qquad (3.3.1)$$

In general $\Delta(g,s)$ will be reducible to the form

$$\Delta(g,s) = \Delta_1(g,s)\Delta_2(g,s) \ldots\ldots \Delta_\ell(g,s) \qquad (3.3.2)$$

where the factors $\{\Delta_i(g,s):i=1,2,\ldots,\ell\}$ are polynomials in
g which are irreducible over the field of rational functions
in s.   Let the irreducible factors $\Delta_i(g,s)$ have the form

$$\Delta_i(g,s) = g_i^{t_i} + a_{i1}(s)g_i^{t_i-1} + \ldots\ldots + a_{it_i}(s) = 0 \qquad (3.3.3)$$

where t_i is the degree of the ith irreducible polynomial
and the coefficients $\{a_{ij}(s):i=1,2,\ldots,\ell;j=1,2,\ldots,t_i\}$ are
rational functions in s.   Then if $b_{io}(s)$ is the least
common denominator of the coefficients $\{a_{ij}(s):j=1,2,\ldots,t_i\}$
equation (3.3.3) can be put in the form

$$b_{io}(s)g_i^{t_i} + b_{i1}(s)g_i^{t_i-1} + \ldots\ldots + b_{it_i}(s) = 0 \quad (3.3.4)$$

$$i = 1, 2, \ldots, \ell$$

where the coefficients $\{b_{ij}(s): i=1,2,\ldots,\ell; j=1,2,\ldots,t_i\}$ are polynomials in s. The function of a complex variable $g_i(s)$ defined by equation (3.3.4) is called an algebraic function [1; appendix 1]. Thus associated with an open-loop gain matrix G(s) is a set of algebraic functions $\{g_i(s): i=1,2,\ldots,\ell\}$ which are directly related to the eigenvalues of G(s). The characteristic gain functions of G(s) are defined to be the set of algebraic functions $\{g_i(s):i=1,2,\ldots,\ell\}$.

The problem of finding the irreducible polynomials $\{\Delta_i(g,s):i=1,2,\ldots,\ell\}$ from which the characteristic gain functions are defined is closely linked to the problem of finding an appropriate canonical form of G(s). If $\Delta(g,s)$ was reducible to factors linear in g then G(s) could be put into Jordan form [2]. In general this will not be the case and a suitable canonical form is defined as follows.

Let

$$C(\Delta_i) \triangleq \begin{bmatrix} 0 & 0 & \ldots & 0 & -a_{it_i}(s) \\ 1 & 0 & \ldots & 0 & -a_{i,t_i-1}(s) \\ 0 & 1 & \ldots & 0 & -a_{i,t_i-2}(s) \\ \vdots & & & & \\ 0 & & & 1 & -a_{i1}(s) \end{bmatrix} \quad (3.3.5)$$

for $t_i > 1$ with

$$C(\Delta_i) \triangleq -a_{i1}(s) \qquad \text{if } t_i = 1 \qquad (3.3.6)$$

then a transformation matrix E(s) exists such that

$$G(s) = E(s) \ Q(s) \ E(s)^{-1} \tag{3.3.7}$$

where Q(s) is a unique block diagonal matrix, which is called the <u>irreducible rational canonical form</u> of G(s) and is given by

$$Q(s) \stackrel{\Delta}{=} \text{diag}\left[C(\Delta_1), C(\Delta_2), \ldots\ldots, C(\Delta_\ell)\right] \tag{3.3.8}$$

It is clear that given Q(s) the irreducible factors $\Delta_i(g,s)$ can easily be obtained.   A proposed method for finding Q(s) for any given G(s) is presented in appendix 2.

3.3-1 Poles and zeros of a characteristic gain function

Consider the defining equation for a characteristic gain function g(s) :

$$\Phi(g,s) \stackrel{\Delta}{=} b_o(s)g^t + b_1(s)g^{t-1} + \ldots + b_t(s) = 0 \tag{3.3.9}$$

We will take both

$$b_o(s) \neq 0 \quad \text{and} \quad b_t(s) \neq 0$$

since, if either or both of these polynomial coefficients were to vanish, we could find a reduced-order equation such that both the coefficients of the highest and zeroth powers of g(s) were non-zero; this reduced-order equation would then be taken as defining an appropriate new algebraic function for whose defining equation the supposition would be true.

It may happen however that $b_o(s)$ and $b_t(s)$ share a common factor and thus both vanish together at some specific set of values of s .   Before looking at the effect of this, consider the situation when $b_o(s)$ and $b_t(s)$ do not share a common factor.   The algebraic function will obviously be zero when

$$b_t(s) = 0 \tag{3.3.10}$$

and will tend to infinity as

$$b_o(s) \to 0 \qquad\qquad (3.3.11)$$

For this reason those values of ·s which satisfy equation
(3.3.10) are defined to be the _zeros_ of the algebraic function
g(s), and those values of s which satisfy the equation

$$b_o(s) = 0 \qquad\qquad (3.3.12)$$

are defined to be the _poles_ of the algebraic function g(s).
Unless stated otherwise the terminology 'poles and zeros'
should be taken as referring only to _finite_ poles and zeros.
The point s=∞ requires special attention and is dealt with
at the end of this sub-section.

In order to be able to take equations (3.3.10) and
(3.3.12) as defining the zeros and poles of g(s) in the
general case, we must show that they remain appropriate when
$b_o(s)$ and $b_t(s)$ share a common factor.   Let us first dispose
of the trivial case when all the coefficients $\{b_i(s) : i=0,2,...,t\}$
share a common factor by saying that such a common factor
would simply be divided out to get a new defining equation
for an appropriate algebraic function.   Suppose then that
$b_o(s)$ and $b_t(s)$ have a common factor, but that some non-
empty set of coefficients $\{b_u(s), b_{u+1}(s),..., b_v(s)\}$ do
not share this common factor.   Then dividing through the
left-hand side of equation (3.3.9) by $b_o(s)$ we get

$$g^t + \frac{b_1(s)}{b_o(s)} g^{t-1} + \ldots + \frac{b_u(s)}{b_o(s)} g^{t-u} + \ldots + \frac{b_v(s)}{b_o(s)} g^{t-v} + \ldots + \frac{b_t(s)}{b_o(s)} = 0$$

$$(3.3.13)$$

Then, as $s \to \bar{s}$ where $\bar{s}$ is a zero of the common factor of $b_o(s)$
and $b_t(s)$, the moduli of the coefficient set

$$\{\frac{b_u(s)}{b_o(s)}, \cdot \cdot \cdot , \frac{b_v(s)}{b_o(s)}\}$$

all become arbitrarily large, and it is obvious that $g(s)$
will have a pole at $s = \bar{s}$.

Again, suppose that $b_o(s)$ and $b_t(s)$ have a common
factor but that some non-empty set of coefficients
$\{b_j(s),\ldots,b_m(s)\}$ do not.   Then as $s \rightarrow \bar{s}$ where $\bar{s}$ is
a zero of the common factor, the algebraic equation (3.3.9)
may be replaced by

$$b_j(\bar{s})g^{t-j}(\bar{s}) + \ldots + b_m(\bar{s})g^{t-m}(\bar{s}) = 0 \qquad (3.3.14)$$

where

$$b_j(\bar{s}) \neq 0 , \ldots , b_m(\bar{s}) \neq 0$$

so that we must have

$$g(\bar{s}) = 0$$

showing that $\bar{s}$ is indeed a zero of the algebraic function $g(s)$.

We thus conclude that equations (3.3.10) and (3.3.12)
may be taken as defining the finite zeros and finite poles
of the algebraic function $g(s)$, and that use of these
definitions enables us to cope with the existence of coincident
poles and zeros.   The pole and zero polynomials of $g(s)$,
denoted by $p_g(s)$ and $z_g(s)$, are defined as

$$p_g(s) \triangleq b_o'(s)$$

$$\qquad (3.3.15)$$

and

$$z_g(s) \triangleq b_t'(s)$$

where $b_o'(s)$ and $b_t'(s)$ are the monic polynomials obtained
from $b_o(s)$ and $b_t(s)$ respectively, by dividing each polynomial
by its leading coefficient.

For the purpose of considering $g(s)$ at the point $s=\infty$ we put

$$s=z^{-1} \qquad\qquad (3.3.16)$$

so that

$$\Phi(g,s)=\Phi(g,z^{-1})=z^{-q}\ \Psi(g,z) \qquad\qquad (3.3.17)$$

where q is the number of finite poles of $g(s)$. In any neighbourhood of the value $z=0$ (the point $z=0$ itself being excluded from it) the equation $\Phi(g,s)=0$ is equivalent to the equation $\Psi(g,z)=0$. Therefore if we consider the equation

$$\Psi(g,z)\overset{\Delta}{=}c_o(z)g^t+c_1(z)g^{t-1}+\ldots+c_t(z)=0 \qquad (3.3.18)$$

it follows that:

(i) $s=\infty$ is a pole of the characteristic gain function $g(s)$ if and only if $c_o(o)=0$

(ii) $s=\infty$ is a zero of the characteristic gain function $g(s)$ if and only if $c_t(o)=0$

For an open-loop gain matrix $G(s)$ describing a physically realizable system, which by definition (see section 2.1) we are considering here, it is not possible for $g(s)$ to have poles at infinity. In fact it is easy to show that for $s=\infty$ the values of the characteristic gain function $g(s)$ are simply the eigenvalues of D.

3.3-2   Algebraic definition of poles and zeros for a transfer function matrix

Let $T(s)$ be an $m\times\ell$ rational matrix-valued function of the complex variable s. Then there exists a canonical form for $T(s)$, the Smith-McMillan form $\begin{bmatrix}3\end{bmatrix}$ $M(s)$, such that

$$T(s) = H(s)M(s)J(s) \qquad\qquad (3.3.19)$$

where the m×m matrix H(s) and the ℓ×ℓ matrix J(s) are both
unimodular (that is having a constant value for their determinants,
independent of s).   If r is the normal rank of T(s)
(that is T(s) has rank r for almost all values of s) then
M(s) has the form

$$M(s) = \begin{bmatrix} M^*(s)_{rr} & O_{r,\ell-r} \\ O_{m-r,r} & O_{m-r,m-r} \end{bmatrix} \qquad (3.3.20)$$

with

$$M^*(s) = \text{diag} \left[\frac{\varepsilon_1(s)}{\psi_1(s)} , \frac{\varepsilon_2(s)}{\psi_2(s)} , \dots , \frac{\varepsilon_r(s)}{\psi_r(s)} \right] (3.3.21)$$

where:

    (i)   each $\varepsilon_i(s)$ divides all $\varepsilon_{i+j}(s)$ and

    (ii)   each $\psi_i(s)$ divides all $\psi_{i-j}(s)$.

With an appropriate partitioning of H(s),M(s) and J(s) we
therefore have

$$T(s) = \begin{bmatrix} H_1(s) & H_2(s) \end{bmatrix} \begin{bmatrix} M^*(s) & O \\ O & O \end{bmatrix} \begin{bmatrix} J_1(s) \\ J_2(s) \end{bmatrix}$$

$$= H_1(s)M^*(s)J_1(s) \qquad (3.3.22)$$

where M*(s) is as defined in equation (3.3.21).

Thus T(s) may be expressed in the form

$$T(s) = H_1(s) \left[\text{diag}\{ \frac{\varepsilon_i(s)}{\psi_i(s)} \} \right] J_1(s)$$

$$= \sum_{i=1}^{r} h_i(s) \frac{\varepsilon_i(s)}{\psi_i(s)} j_i^t(s) \qquad (3.3.23)$$

where:

    (i) $\{h_i(s) : i = 1,2,\dots,r\}$ are the columns of the
matrix $H_1(s)$;

(ii) $\{j_i^t(s) : i = 1,2,\ldots,r\}$ are the rows of the
matrix $J_1(s)$.

We know that

$$r \leqslant \min(\ell,m)$$

and that $H(s)$ and $J(s)$ are unimodular matrices of full
rank m and ℓ respectively for all s.   Suppose $T(s)$ is the
transfer function matrix for a system with input transform
vector $\hat{u}(s)$ and output transform vector $\hat{y}(s)$.   Then any
input vector $\hat{u}(s)$ is turned into an output vector $\hat{y}(s)$
by

$$\hat{y}(s) \; = \; \sum_{i=1}^{r} \; h_i(s) \; \frac{\varepsilon_i(s)}{\psi_i(s)} \; \left[j_i^t(s)\hat{u}(s) \right] \qquad (3.3.24)$$

For the single-input single-output case where

$$\hat{y}(s) \; = \; \frac{k\varepsilon(s)}{\psi(s)} \qquad \hat{u}(s)$$

with   k   a constant, the transfer function

$$g(s) \; = \; \frac{k\varepsilon(s)}{\psi(s)}$$

is defined as having zeros at those values of s where $\varepsilon(s)$
vanishes and poles at those values of s where $\psi(s)$ vanishes.
Thus for a non-zero $\hat{u}(s)$ the modulus of $\hat{y}(s)$ vanishes
when s is a zero of $g(s)$, and becomes arbitrarily large when s
is a pole of $g(s)$.   A natural way therefore to characterize
the zeros and poles of $T(s)$ is in terms of those values of
s for which $\| \hat{y}(s) \|$ becomes zero for non-zero $\| \hat{u}(s) \|$,
and arbitrarily large for finite $\| \hat{u}(s) \|$, where $\| \cdot \|$
denotes the standard vector norm.   This natural extension

of scalar case ideas leads directly to definitions of zeros
and poles of T(s) in terms of the Smith-McMillan form
quantities

$$\{\frac{\varepsilon_i(s)}{\psi_i(s)}\}$$

because of the following pair of simple results.

<u>Zero lemma</u>: $\|\hat{y}(s)\|$ vanishes for $\|\hat{u}(s)\| \neq 0$ and s finite
if and only if some $\varepsilon_i(s)$ is zero.

<u>Pole lemma</u>: $\|\hat{y}(s)\| \rightarrow \infty$ for $\|\hat{u}(s)\| < \infty$ if and only if
some $\psi_i(s) \rightarrow 0$.

These considerations lead naturally to the following
definitions [3].

<u>Poles of T(s)</u>: The poles of T(s) are defined to be the set
of all zeros of the set of polynomials $\{\psi_i(s) : i = 1,2,\ldots,r\}$.
In what follows we will usually denote the poles of T(s) by
$\{p_1, p_2, \ldots, p_n\}$ and put

$$p_T(s) = (s-p_1)(s-p_2) \ldots (s-p_n) \qquad (3.3.25)$$

where $p_T(s)$ is conveniently referred to as the pole polynomial
of T(s) and is given by

$$p_T(s) = \prod_{i=1}^{r} \psi_i(s) \qquad (3.3.26)$$

<u>Zeros of T(s)</u>: The zeros of T(s) are defined to be the set
of all zeros of the set of polynomials $\{\varepsilon_i(s) : i = 1,2,\ldots,r\}$.
We will normally denote the zeros of T(s) by $\{z_1, z_2, \ldots, z_\omega\}$

and put

$$z_T(s) = (s-z_1)(s-z_2) \ldots (s-z_\omega) \qquad (3.3.27)$$

where $z_T(s)$ is conveniently referred to as the zero polynomial of T(s) and is given by

$$z_T(s) = \prod_{i=1}^{r} \varepsilon_i(s) \qquad\qquad (3.3.28)$$

It is important to remember that $z_T(s)$ and $p_T(s)$ are not necessarily relatively prime; for this reason it is wrong to simply define $z_T(s)$ and $p_T(s)$ for a square matrix T(s) as the numerator and demominator polynomials of det T(s).

Rules for calculating pole polynomials and zero polynomials

The route via the Smith-McMillan form is not always convenient for the determination of the poles and zeros of T(s), particularly if the calculation is being done by hand. The following rules [4] can be shown to give the same results as the Smith-McMillan definitions.

Pole polynomial rule: $p_T(s)$ is the monic polynomial obtained from the least common denominator of all non-zero minors of all orders of T(s).

Zero polynomial rule: $z_T(s)$ is the monic polynomial obtained from the greatest common divisor of the numerators of all minors of T(s) of order r (r being the normal rank of T(s)) which minors have all been adjusted to have $p_T(s)$ as their common denominator

3.3-3 Relationship between algebraically defined poles/zeros of the open-loop gain matrix G(s) and the poles/zeros of the corresponding set of characteristic gain functions

As a key step in the establishment of a generalized Nyquist stability criterion, it is crucially important to relate the poles and zeros defined by algebraic means to complex variable theory, and thus to the poles and zeros of the set of characteristic gain functions.

The coefficients $a_i(s)$ in the expansion

$$\det \left[gI_m - G(s) \right] = g^m + a_1(s)g^{m-1} + a_2(s)g^{m-2} + \ldots + a_m(s)$$

$$(3.3.29)$$

are all appropriate sums of minors of $Q(s)$ since it is well known that:

$$\det \left[gI_m - G(s) \right]$$

$$= g^m - \left[\text{trace } G(s) \right] g^{m-1} + \left[\Sigma \text{principal minors of } G(s) \text{ of order } 2 \right] g^{m-2}$$

$$- \ldots + (-1)^m \det G(s) \qquad (3.3.30)$$

and thus the pole polynomial $b_o'(s)$ is the monic polynomial obtained from the least common denominator of all non-zero principal minors of all orders of $G(s)$.

Now the pole polynomial $p_G(s)$ of a square matrix $G(s)$

is the monic polynomial obtained from the least common denominator of all non-zero minors of all orders of $G(s)$. Therefore, if $e_G(s)$ is the monic polynomial obtained from the least common denominator of all non-zero non-principal minors, with all factors common to $b_o'(s)$ removed, we have that

$$p_G(s) = e_G(s)b_o'(s) \qquad (3.3.31)$$

Furthermore since

$$\det G(s) = a_m(s) = \frac{b_m(s)}{b_o(s)} \qquad (3.3.32)$$

and since from the Smith-McMillan form for $G(s)$

$$\det G(s) = \alpha. \frac{z_G(s)}{p_G(s)} \qquad (3.3.33)$$

where α is a scalar quantity independent of s, we must have that

$$z_G(s) \quad = \quad e_G(s)b_m^{'}(s) \tag{3.3.34}$$

In many cases the least common denominator of the non-zero non-principal minors of G(s) will divide $b_o(s)$, in which case $e_G(s)$ will be unity and the pole and zero polynomials for G(s) will be $b_o^{'}(s)$ and $b_m^{'}(s)$ respectively. In general a square-matrix-valued function of a complex variable G(s) will have a set of ℓ irreducible characteristic gain functions in the form specified by equation (3.3.3) and the general form for thepole and zero polynomials can be written as

$$P_G(s) \quad = \quad e_G(s) \prod_{i=1}^{\ell} b_{io}^{'}(s) \tag{3.3.35}$$

and

$$z_G(s) \quad = \quad e_G(s) \prod_{i=1}^{\ell} b_{i,t_i}^{'}(s) \tag{3.3.36}$$

where the pole and zero polynomials for the jth characteristic gain function $g_j(s)$ are $b_{jo}^{'}(s)$ and $b_{j,t_j}^{'}(s)$ respectively.

Example demonstrating the pole-zero relationships

Let

$$G(s) \quad = \quad \frac{1}{(s+1)(s+2)(s-1)} \begin{bmatrix} (s-1)(s+2) & 0 \\ -(s+1)(s+2) & (s-1)(s+1) \end{bmatrix}$$

$$= \begin{bmatrix} \frac{1}{s+1} & 0 \\ \frac{-1}{s-1} & \frac{1}{s+2} \end{bmatrix}$$

The pole polynomial for G(s) is obviously

$$P_G(s) \quad = \quad (s+1)(s+2)(s-1)$$

and consequently the zero polynomial is

$$z_G(s) \quad = \quad (s-1) \ .$$

The characteristic equation for G(s) is

$$\det \left[gI - G(s) \right] \;=\; \left(g - \frac{1}{s+1} \right) \left(g - \frac{1}{s+2} \right) \;=\; 0$$

so that the irreducible characteristic equations are

$$\Delta_1(g,s) \;=\; g - \frac{1}{s+1} \;=\; 0$$

and

$$\Delta_2(g,s) \;=\; g - \frac{1}{s+2} \;=\; 0$$

which may be written as

$$(s+1)g - 1 \;=\; 0$$

and

$$(s+2)g - 1 \;=\; 0 \;.$$

Therefore the pole and zero polynomials for the characteristic gain functions $g_1(s)$ and $g_2(s)$ are

$$p_{g_1}(s) \;=\; b_{10}(s) \;=\; (s+1) \qquad z_{g_1}(s) = \frac{b_{11}(s)}{-1} \;=\; 1$$

$$p_{g_2}(s) \;=\; b_{20}(s) \;=\; (s+2) \qquad z_{g_2}(s) = \frac{b_{21}(s)}{-1} \;=\; 1$$

Now for G(s) the monic polynomial obtained from the least common denominator of all non-zero non-principal minors with all factors common to $b_o'(s)$ $(=b_{10}'(s)b_{20}'(s))$ removed is given by

$$e_G(s) \;=\; (s-1)$$

which verifies the relationships

$$p_G(s) \;=\; e_G(s) \prod_{i=1}^{2} b_{io}'(s)$$

and

$$z_G(s) \;=\; e_G(s) \prod_{i=1}^{2} b_{i1}'(s)$$

3.3-4 Riemann surface of a characteristic gain function

A characteristic gain function g(s) is defined by an irreducible equation of the form

$$b_o(s)g^t + b_1(s)g^{t-1} + \ldots + b_t(s) = 0 \quad (3.3.37)$$

having in general t distinct finite roots.   An exception

occurs only if

(a) $b_o(s) = 0$, because the degree of the equation is then

lowered, and as $b_o(s) \rightarrow 0$ one or more of the roots becomes

infinite;   or if

(b)   the equation has multiple roots.

This last situation can occur for finite values of s if, and

only if, an expression, called the discriminant of the

equation, vanishes.   The discriminant [5] is an entire rational

function of the equation coefficients;   it will be denoted

by $D_g(s)$, and is discussed in appendix 3.

Ordinary points of the characteristic gain function

An ordinary point [1;6] of the characteristic gain

function g(s) is any finite point of the complex plane such

that $b_o(s) \neq 0$ and $D_g(s) \neq 0$.

Critical points of the characteristic gain function

A critical point [1; 6] of g(s) is any point of the

complex plane at which either

$$b_o(s) = 0 \quad \text{or} \quad D_g(s) = 0,$$

or both, plus the point $s=\infty$.

Branch points of the characteristic function

Solutions of

$$D_g(s) = 0$$

are called finite branch points of the characteristic gain

function.   The point at infinity is a branch point if the

discriminant $\bar{D}_g(z)$ of equation (3.3.18) satisfies $\bar{D}_g(0) = 0$.

At every ordinary point the equation (3.3.37) defining the

characteristic gain function has t distinct roots, since
the discriminant does not vanish.   The theory of algebraic
functions [1] then shows that in a simply connected region
of the complex plane punctured by the exclusion of the critical
points the values of the characteristic gain function g(s)
form a set of analytic functions;   each of these analytic
functions is called a branch of the characteristic gain
function g(s).   Arguments based on standard techniques of
analytic continuation, together with the properties of
algebraic equations, show that the various branches can be
organized into a single entity: the corresponding algebraic
function.   This is summarized in the following basic theorem
of algebraic function theory: an irreducible algebraic
equation of the form (3.3.37) defines precisely one t-valued
regular function g(s) in the punctured plane [7] .

Functions defined in this way are called algebraic
functions, and can be regarded as natural generalizations
of the familiar elementary functions of a complex variable.
An elementary function of a complex variable has the set of
complex numbers $\mathbb{C}$ as both its domain and its range.   An
algebraic function has the complex number set $\mathbb{C}$ as its range
but has a new and appropriately defined domain $\mathbb{R}$ which is
called its   Riemann Surface [8] .   Since the Riemann
surface of an algebraic function plays a crucial role in this
work   it is important to have an intuitive grasp of the ideas
underlying its definition and formation, which is therefore
now briefly considered.

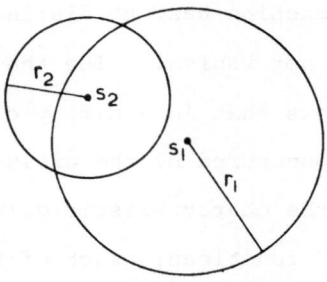

Figure 6. Analytic continuation

Suppose we have a representation of part of one branch of
an algebraic function in the form of a power series; such
a representation is usually called a functional element.
Imagine its circle of convergence to be cut out of paper and
that the individual points of the paper disc are made bearers
of the unique functional values of the elements. If now this
initital element is analytically continued by means of a
second power series, another circle of convergence can be
thought of as being cut out and pasted partly over the first,
as illustrated by figure 6. The parts pasted together are
made bearers of the same functional values and are accordingly
treated as a single region covered once with values. If a
further analytic continuation is carried out, a further disc
is similarly pasted on to the preceding one. Now suppose
that, after repeated analytic continuations, one of the discs
lies over another disc, not associated with an immediately

preceding analytic continuation, as shown in figure 7.
Such an overlapping disc is pasted together with the one
it overlaps if and only if both are bearers of the same
functional values.   If,however,they bear different functional
values they are allowed to overlap but remain disconnected.
Thus two sheets, which are bearers of different functional
values, become superimposed on this part of the complex plane.

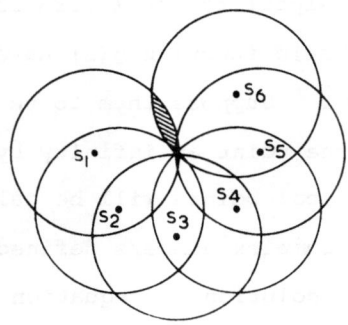

Figure 7. Repeated analytic continuation

Continuing this process for as long as possible, a
surface-like configuration is obtained covering t "sheets"
of the complex plane, where t is the degree of the algebraic
function.   To form the Riemann surface these sheets can be
joined together in the most varied of ways.   This may
involve connecting together two sheets which are separated
by several other sheets lying between them.   Although such
a construction cannot be carried out in a three-dimensional
space it is not difficult to give a perfectly satisfactory
topological description of the process required.   This
surface-like configuration is called the Riemann surface
of the multiple-valued algebraic function.   On the Riemann

surface the entire domain of values of the algebraic
function is spread out in a completely single-valued manner
so that, on every one of the t copies of the complex plane
involved, every point is the bearer of one and only one
value of the function.

A method for building Riemann surfaces is given in
appendix 4. This involves the use of cuts in the complex
plane and it may be helpful to say a word about them at this
point. Let an algebraic function g(s) have r critical
points $\{ a_1, a_2, \ldots, a_r \}$. Suppose them to be joined to one
another and then to the point at infinity by a line L.
Any line joining critical points will be called a cut. Let
$\mathbb{L}$ denote the set of complex numbers defined by the line L.
We then have that the solutions of equation (3.3.37) define
a set of t "distinct" analytic functions $\{ \bar{g}_1(s), \bar{g}_2(s), \ldots, \bar{g}_t(s) \}$
in the cut plane $\mathbb{C} - \mathbb{L}$. Each of these functions can be
analytically continued, by standard procedures, across the
cut L. Now it follows from the fundamental principles
of analytic continuation that if an analytic function
satisfies an algebraic equation in one part of its domain
of definition, it must satisfy that equation in every region
into which it is analytically continued. We must therefore
have that:

(i) there are only t "distinct" analytic functions
which satisfy the defining algebraic equation in the cut plane
$\mathbb{C} - \mathbb{L}$;

(ii) each analytic continuation of any of these
analytic functions $\{ \bar{g}_i(s) : i = 1, 2, \ldots, t \}$ gives rise to an

analytic function which also satisfies the defining algebraic
equation.   It follows from this that the set of analytic
functions associated with one side of the cut L must be a
simple permutation of the set of analytic functions
associated with the other side of the cut.   Therefore by
identifying and suitably matching up corresponding analytic
functions (via their sets of computed values) on opposite
sides of the cut L , one can produce an appropriate domain
on which a single analytic function may be specified which
defines a continuous single-valued mapping from this domain
into the complex plane.   This function is of course the
algebraic function, conceived of as a single entity, and
the domain so constructed is its Riemann surface.

It is sufficient for the purposes of understanding this
book for the reader to know that a Riemann surface can be
constructed for any given algebraic function, on which its
values form a single-valued function of position.   Many
standard relationships and properties of analytic function
theory generalize, using the Riemann surface concept, to the
algebraic function case and, in particular the Principle of
the Argument holds on the Riemann surface;   an extension of
the Principle of the Argument is developed in appendix 5.

The Riemann surface which is the domain of the
characteristic gain function g(s) will be called the
frequency surface or s-surface.   When the open-loop gain
matrix G(s) is mxm and has a corresponding characteristic
equation which is irreducible (i.e. the usual case in practice)
the frequency surface is formed out of m copies of the complex

frequency plane or s-plane.

3.3-5   Generalized root locus diagrams

The characteristic gain function g(s) is a function of
a complex variable whose poles and zeros are located on the
frequency surface domain.   It is convenient to exhibit the
nature of g(s) by drawing constant phase and constant
magnitude contours of g(s) on the frequency surface.   If
the computational method outlined in appendix 4 is used to
construct the surface then the superposition of constant
phase and magnitude contours is clearly a simple process.
The frequency surface can be thought of as the set of all
possible closed-loop characteristic frequencies associated
with all possible values of the complex gain parameter g.
When the surface is characterized by constant phase and
magnitude contours of g(s) we have a direct correspondence
between a closed-loop characteristic frequency and an open-
loop gain, and since the surface is constructed from m
copies of the complex frequency plane, for each value of s
there are m corresponding characteristic gains.

From equation (3.1.4) we have

$$g(s) = -\frac{1}{k} \qquad\qquad (3.3.38)$$

so that the variation of the closed-loop poles (characteristic
frequencies) with the real control variable k traces out
loci which are equivalent to the 180° phase contours of g(s).
Equation (3.3.38) is a direct generalization of the defining
equation for the single-loop root locus diagram.   The 180°
phase contours of g(s) are the multivariable root loci i.e.
the variation of the closed-loop poles with the gain control
variable k.   The fact that multivariable root loci 'live'

on a Riemann surface explains their complicated behaviour
[9] as compared with the single-input, single-output case
where the root loci lie on a simple complex plane (a trivial,
i.e. one sheeted, Riemann surface). The multivariable
root loci will sometimes be referred to as the characteristic
frequency loci.

Recall that in section 3.2 it was pointed out that the
characteristic equations for G(s) and S(g) are in general
different in that the equation for S(g) may contain factors
of s which are independent of g. These factors therefore
correspond to closed-loop poles which are independent of g,
or equivalently independent of the gain control variable k;
and, from the root locus point of view, these factors
correspond to degenerate loci each consisting of a single
point. The degenerate loci are therefore not picked out by
the 180° phase contours of g(s) on the frequency surface.
In practice the characteristic frequency loci are generated
as the set of loci in a single copy of the complex frequency
plane traced out by the eigenvalues of S(g) as g traverses
the negative real axis in the gain plane. This approach
automatically picks out the degenerate loci. In common
with the classical root locus approach of Evans the characteristic
frequency loci are usually calibrated in terms of the gain
control variable $k=-g^{-1}$.

3.3-6   Example of frequency surface and characteristic
frequency loci

As an illustrative example consider the general multivariable feedback configuration of figure 3 with a corresponding
open-loop gain matrix

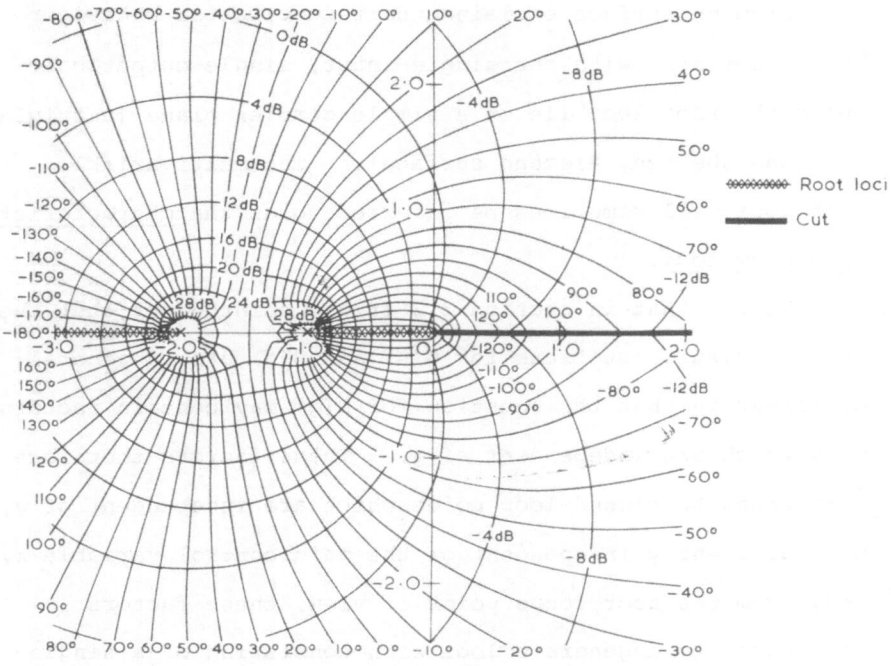

Figure 8. Sheet 1 of the frequency surface

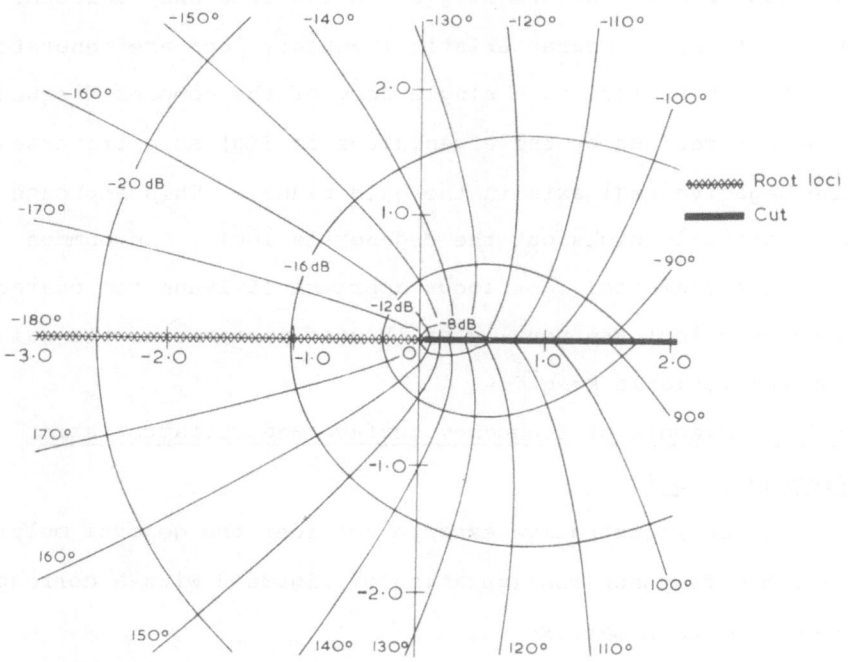

Figure 9. Sheet 2 of the frequency surface

$$G(s) = \frac{1}{1.25(s+1)(s+2)} \begin{bmatrix} s-1 & s \\ -6 & s-2 \end{bmatrix}$$

The matrix is of order two and therefore the appropriate surface will be constructed from two sheets of the complex s-plane. The two sheets are shown characterized by constant phase and magnitude contours of g(s) in figures 8 and 9. The cuts, identifiable by discontinuities in the contours, are represented by thick black lines; and the characteristic frequency loci, which are the 180° phase contours of g(s), are identified by a diamond symbol. The characteristic frequency loci indicate that variation of the gain control parameter k, upwards from zero, causes the system to experience stability, instability and stability again. This phenomenon is clearly linked with the presence of a branch point in the right half-plane (at $s=\frac{1}{24}$).

Note that since we have completely characterized the feedback configuration by its open-loop gain matrix there are no unobservable or uncontrollable modes.

3.4   Characteristic frequency functions

The natural way to define the characteristic frequency function s(g) is via the characteristic equation

$$\nabla(s,g) \triangleq \det\left[sI_n - S(g)\right] = 0 \qquad (3.4.1)$$

It is an algebraic function and the detailed study of the characteristic gain function presented in the previous section can be applied directly to it with the roles of s and g reversed.

The Riemann surface which is the domain of the characteristic frequency function will be called the gain surface or g-surface.   It is formed out of n copies of the complex gain plane or g-plane since there are n values of closed-loop characteristic frequency (closed-loop poles) for every value of g.   The gain surface can be thought of as the set of all possible open-loop characteristic gains of the open-loop gain matrix G(s) associated with all possible closed-loop characteristic frequencies.   In a similar fashion to the gain function g(s) it is convenient to exhibit the behaviour of s(g) on the gain surface by superimposing constant phase and magnitude contours of s(g) onto the surface.   Like g(s) the frequency function s(g) has poles and zeros but their significance is quite different.

3.4-1 Generalized Nyquist diagram

Each 'sheet' of a gain surface characterized by constant phase and magnitude contours of s(g)is divided into regions corresponding to left half-plane and right half-plane closed -loop characteristic frequencies.   Therefore given such a calibrated surface one can see at a glance which values of g (or equivalently k) correspond to stable closed-loop poles.   The boundary between stable and unstable regions is clearly the $\pm 90^{\circ}$ phase contours of s(g).   The $\pm 90^{\circ}$ phase contours of s(g) are a natural generalization of the single-loop Nyquist diagram and are called characteristic gain loci.

In practice the characteristic gain loci are generated as the loci in the complex gain plane traced out by the eigenvalues of G(s) as s traverses the so called Nyquist D-contour in the s-plane.   Suppose that we consider a portion of the imaginary axis.   We can then compute a set of loci corresponding to the eigenvalues $\bar{g}_1(j\omega),\ldots,\bar{g}_m(j\omega)$ (where in this context $j = \sqrt{-1}$)

in the following way:

(i)   Select a value of angular frequency, say ω_a .

(ii)   Compute the complex matrix $G(j\omega_a)$.

(iii) Use a standard computer algorithm to compute the
       eigenvalues of $G(j\omega_a)$, which are a set of complex
       numbers denoted by $\{\bar{g}_i(j\omega_a)\}$.

(iv)   Plot the numbers $\{\bar{g}_i(j\omega_a)\}$ in the complex plane.

(v)   Repeat with further values of angular frequency $\omega_b, \omega_c,$
       ... etc., and join the resulting plots up into
       continuous loci using a sorting routine based on the
       continuity of the various branches of the characteristic
       functions involved.

For the purpose of developing a generalized Nyquist
stability criterion in chapter 4 the Nyquist D-contour is
traversed in the standard <u>clockwise</u> direction.

3.4-2   Example of gain surface and characteristic gain loci

As an illustrative example consider the open-loop
gain matrix considered in subsection 3.3-5 which has a
minimal state-space realization

$$A = \begin{bmatrix} -2 & 0 \\ 0 & -1 \end{bmatrix} \qquad B = \begin{bmatrix} 1 & 0.6 \\ 1 & 0.5 \end{bmatrix}$$

$$C = \begin{bmatrix} 2.4 & -1.6 \\ 4.8 & -4.8 \end{bmatrix} \qquad D = \begin{bmatrix} 0 & 0 \\ 0 & 0 \end{bmatrix}$$

The system has two states and therefore the appropriate
gain surface will be constructed from two sheets of the
complex g-plane. The two sheets are shown characterized
by constant phase and magnitude contours of $s(g)$ in figures
10 and 11.   The characteristic gain loci, which are the
$\pm 90^\circ$ phase contours of $s(g)$, are denoted by a series of
crosses.

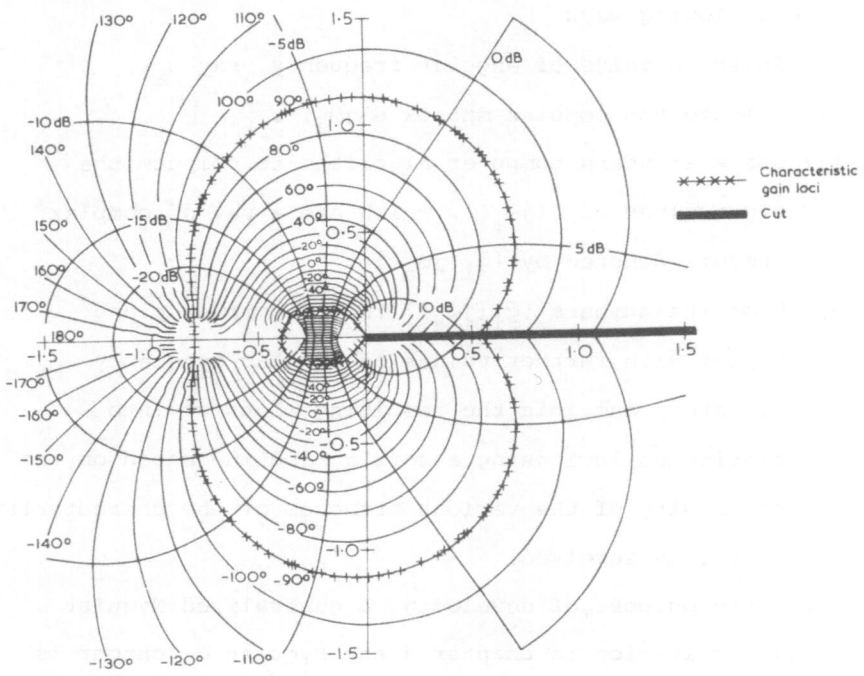

Figure 10. Sheet 1 of the gain surface

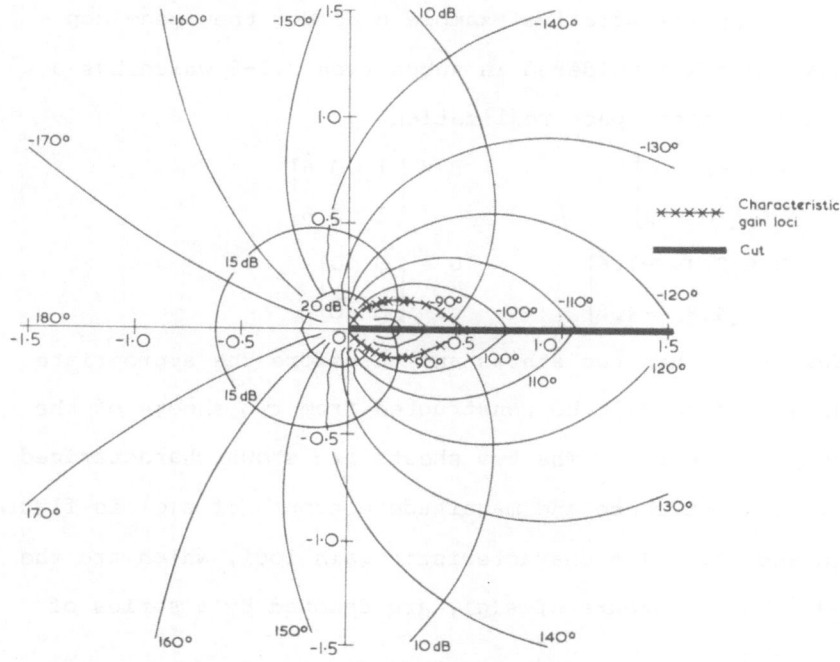

Figure 11. Sheet 2 of the gain surface

53

Right half
plane region

Left half
plane region

Characteristic
gain loci

Cut between
branch points.

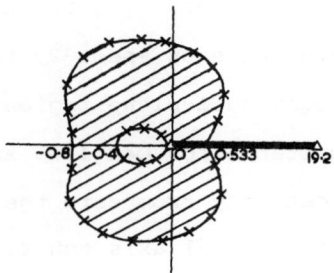

Figure 12. Sketch of figure 10 emphasizing right
half and left half-plane regions

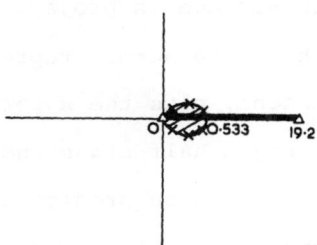

Figure 13. Sketch of figure 11 emphasizing right
half and left half-plane regions

In figures 12 and 13 sketches of figures 10 and 11 are made to emphasize the right and left half-plane regions of closed-loop poles. From these sketches it is easy to infer bounds on the gain control parameter k for stability. Since $g=-k^{-1}$, as we increase k positively from zero the critical value of g moves from $-\infty$ along the real axis towards the origin on each sheet. On sheet 1, g is in a right half-plane region for 1.25<k<2.5 while on sheet 2, for positive k, g never moves into a right half-plane region. Therefore the closed-loop system is stable for 0≤k<1.25 and 2.5<k<∞.

If k is increased negatively from zero the critical value of g moves from ∞ along the real axis towards the origin on each sheet. On both sheets the value of g is in right half-plane regions for $-\infty$<k≤-1.875. Therefore for negative k, which corresponds to positive feedback, the closed-loop system is stable for -1.875<k≤0.

If the calibrated gain surface is projected onto the complex gain plane $\mathbb{C}$, we have the normal representation of the characteristic gain loci, plus the superposition of contours representing both right half-plane and left half-plane regions. Stability can now be predicted by considering the right half-plane regions in relation to a single critical point $(-\frac{1}{k}+j0)$. However, this presentation will in general be difficult to comprehend because of the overlapping of contours. Therefore, although the counting of encirclements in the generalized Nyquist stability criterion [chapter 4] is not fundamental to system stability as pointed out by Saeks [10], it does afford the simplest method of predicting closed-loop stability in the gain plane $\mathbb{C}$.

From a gain surface plot it is possible to determine the closed-loop poles and hence the relative stability of a closed-loop system. This is now illustrated by finding the dominant closed-loop poles for the example under consideration with unity k. It is convenient for this purpose to have the gain surface characterized by constant real and imaginary contours as shown in figures 14 and 15.

Let the dominant closed-loop poles be

$$s_d = \alpha \pm j\beta$$

Then α is the smallest (in magnitude) negative real contour that passes through any one of the critical, -1, values of g, and β is the corresponding imaginary contour. From figures 14 and 15 we have

$$\alpha \simeq -0.05 \text{ and } \beta = 0$$

so that

$$s_d \simeq -0.05$$

By hand calculation the dominant closed-loop pole is

$$s_d = -0.0528$$

Also from a gain surface it is possible to determine the characteristic frequency loci. Apart from possible single-point loci, the root loci are simply the values of s at which the characteristic gain loci has a phase of 180°. Therefore from a gain surface plot the root loci are determined by the values of the constant contours as they cross the negative real axis on each sheet. Similarly, from a calibrated frequency surface, it is possible to determine the characteristic gain loci from the values of constant contours as they cross the imaginary axes.

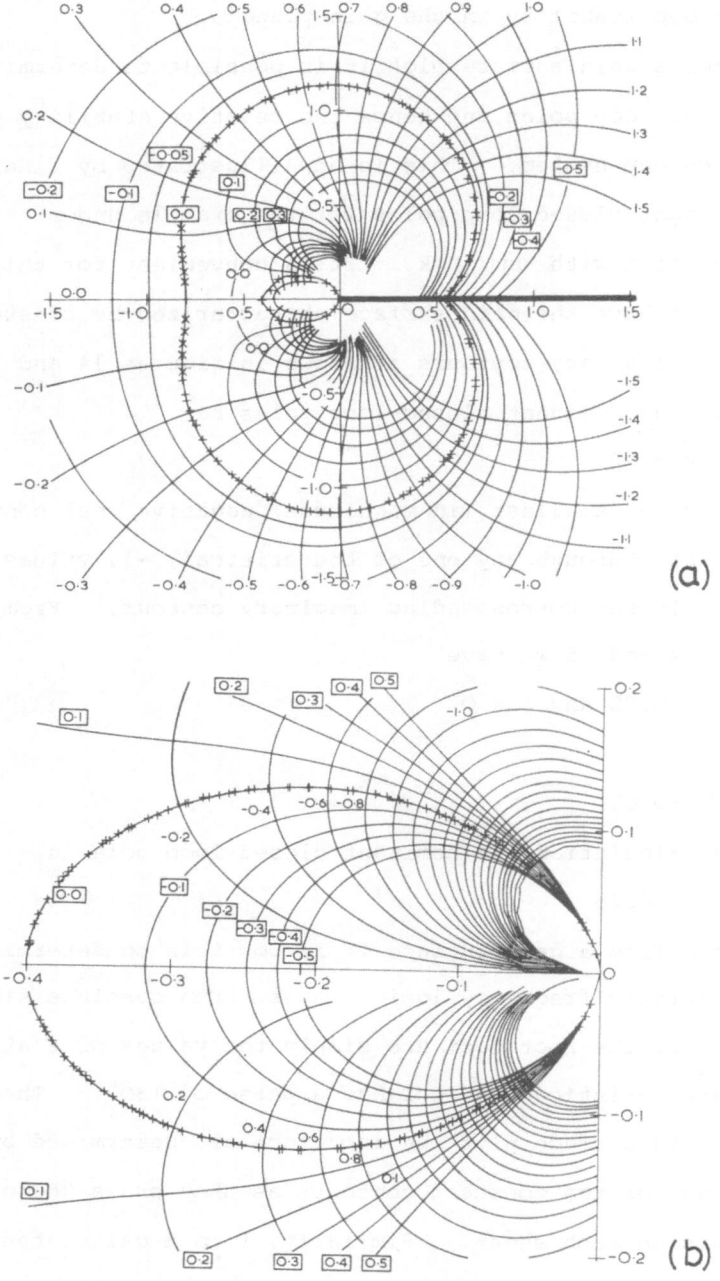

Figure 14. Sheet 1 of the gain surface (a) 'complete' sheet
(b) small region about −0·2

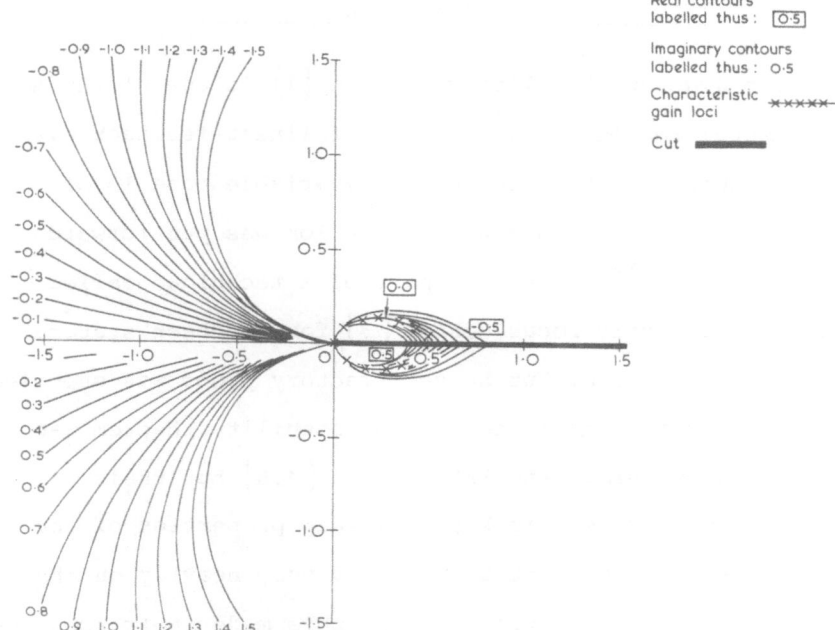

Figure 15. Sheet 2 of the gain surface

References

[1] G.A. Bliss, "Algebraic Functions", Dover, New York, 1966 (reprint of 1933 original).

[2] P.M. Cohn, "Algebra", vol. 1, Wiley, London 1974.

[3] H.H. Rosenbrock, "State Space and Multivariable Theory", Nelson, London, 1970.

[4] T. Kontakos, Ph.D. Thesis, University of Manchester, 1973.

[5] S. Barnett, "Matrices in Control Theory", Van Nostrand-Reinhold, London, 1971.

[6] E. Hille, "Analytic Function Theory", Vol. 2, Ginn and Co., U.S.A., 1962.

[7] K. Knopp "Theory of Functions", Part 2, Dover, New York, 1947.

[8] G. Springer, "Introduction to Riemann Surfaces", Addison-Wesley, Reading, Mass., 1957.

[9] B. Kouvaritakis and U. Shaked, "Asymptotic behaviour of root-loci of multivariable systems", Int. J. Control, 23, 297-340, 1977.

[10] R. Saeks, "On the Encirclement Condition and Its Generalization", IEEE Trans. on Circuits and Systems, 22, 780-785, 1975.

4.   A generalized Nyquist stability criterion

The Nyquist stability criterion [1] is one of the most fundamental results in the theory of linear feedback systems and its generalization to the multivariable case is of great interest.   Such a generalization was put forward by MacFarlane[2] and used as part of a technique called the Characteristic Locus Method [3] for feedback systems analysis and design, but no satisfactory proof was supplied. The proof of a generalized Nyquist stability theorem was undertaken by Barman and Katzenelson [4;5] but their approach ignored certain key algebraic properties of the quantities involved; it also leaned very heavily on the use of cuts in the complex plane; this made their treatment technically complicated, and obscured the essential simplicity of the result.   The purpose of this chapter is to give a rigorous proof of a generalized Nyquist-like stability criterion for the general feedback configuration based on a fundamental result in complex variable theory: the Principle of the Argument applied to an algebraic function defined on an appropriate Riemann surface [appendix 5].

Two Nyquist-like stability tests are in fact stated and proved, the usefulness of each depending on how the subsystems are characterized.   The two statements of the criterion are given in section 4.1 and proved in section 4.2.

4.1   Generalized Nyquist stability criterion

If the subsystems of the general feedback configuration of figure 3 are each characterized by a state-space model then the following statement of the criterion is applicable.

Statement 1. The general feedback configuration is closed-
loop stable if and only if:

(1)   the net sum of anti-clockwise encirclements of the
critical point $(-\frac{1}{k}+j0)$ by the set of characteristic gain
loci is equal to the number of right half-plane poles of G(s);

(2)   the characteristic gain loci do not pass through the
critical point $(-\frac{1}{k}+j0)$;

(3)   the number of branches of the characteristic gain loci
passing through infinity is equal to the number of poles
of G(s) on the imaginary axis; and

(4)   the eigenvalues of the A-matrix, of the open-loop
system S(A,B,C,D), which correspond to modes of the system
which are unobservable and/or uncontrollable from the point
of view of considering the input as that of the first subsystem and the output as that of the hth subsystem, are
all in the left half-plane.

    If the subsystems are completely characterized by
their transfer function matrices, or if it is known that
for each subsystem there are no unobservable and/or
uncontrollable modes in the right half-plane including the
imaginary axis, then the following statement of the criterion
applies.

Statement 2. The general feedback configuration is closed-
loop stable if and only if:

(1)   the net sum of anti-clockwise encirclements of the
critical point $(-\frac{1}{k}+j0)$ by the set of characteristic gain
loci is equal to the total number of right half-plane poles
of $G_1(s), G_2(s), \ldots,$ and $G_h(s)$;

(2)   the characteristic gain loci do not pass through the
critical point $(-\frac{1}{k}+j0)$; and

(3)   the number of branches of the characteristic gain
loci passing through infinity is equal to the number of
poles of G(s) on the imaginary axis.

Note that if condition (2) and/or condition (3) do not
hold the closed-loop system has one or more poles on the
imaginary axis and is therefore not input-output stable
although the equilibrium state at the origin may be stable.

4.2   Proof of the generalized Nyquist stability criterion

In section 2.4 it was shown how the return-difference
operator corresponding to the break point shown in figure 4
is related to the open- and closed-loop characteristic
polynomials by the following expression.

$$\frac{\det F(s)}{\det F(\infty)} = \frac{CLCP(s)}{OLCP(s)} \qquad (4.2.1)$$

This fundamental relationship is the foundation on which
the proof of the generalized Nyquist stability criterion
is based.

The first stage in the proof is to consider the
eigenvalue equation of the return-difference matrix F(s),
that is

$$\det\left[fI_m - F(s)\right] = 0 \qquad (4.2.2)$$

which in general, as for the characteristic equation of
G(s), can be expressed as a product of irreducible algebraic
equations of the form

$$d_{io}(s)f_i^{t_i} + d_{i1}(s)f_i^{t_i-1} + \ldots + d_{it_i}(s) = 0$$
$$i=1,2,\ldots,\ell \qquad (4.2.3)$$

defining a set of algebraic functions $\{f_i(s) : i=1,2,\ldots,\ell\}$

Therefore, as in sub-section 3.3-3 where the poles and zeros of G(s) are related to the poles and zeros of the characteristic gain functions, the pole and zero polynomials of F(s) can be related to the pole and zero polynomials of the algebraic functions $\{f_i(s):i=1,2,\ldots,\ell\}$ as follows

$$p_F(s) = e_F(s) \prod_{i=1}^{\ell} d'_{io}(s)$$

and   (4.2.4)

$$z_F(s) = e_F(s) \prod_{i=1}^{\ell} d'_{it_i}(s)$$

By definition the open-loop characteristic polynomial of the general feedback configuration is given by

$$OLCP(s) \triangleq \det\left[sI_n - A\right]$$
$$= \det\left[sI_{n_1} - A_1\right] \det\left[sI_{n_2} - A_2\right] \cdots \det\left[sI_{n_h} - A_h\right]$$
$$(4.2.5)$$

and it is easily shown [6] that

$$\det\left[sI_{n_i} - A_i\right] = p_{G_i}(s) p_{d_i}(s) \qquad (4.2.6)$$

where $p_{G_i}(s)$ is the pole polynomial for the transfer function matrix $G_i(s)$ and the monic polynomial $p_{d_i}(s)$ has as its zeros the decoupling zeros of the ith subsystem associated with that set of characteristic frequencies (eigenvalues) of A_i which correspond to modes of the ith subsystem which are uncontrollable and/or unobservable. The open-loop characteristic polynomial can therefore be expressed as

$$OLCP(s) = p_{G_1}(s) p_{G_2}(s) \cdots p_{G_h}(s) p_{d_1}(s) p_{d_2}(s) \cdots p_{d_h}(s)$$
$$(4.2.7)$$

The pole polynomial $p_G(s)$ for the open-loop gain matrix G(s) is related to the pole polynomials of the subsystem transfer functions through the relationship

$$p_G(s)p_x(s) \quad = \quad p_{G_1}(s)p_{G_2}(s)\ldots p_{G_h}(s) \qquad (4.2.8)$$

where $p_x(s)$ has as its zeros those poles of $G_1(s),G_2(s),\ldots,$ and $G_h(s)$ which are lost when $G(s)$ is formed $\begin{bmatrix} 7 \end{bmatrix}$. The zeros of $p_x(s)$ are in fact a subset of the unobservable and uncontrollable modes of the system $S(A,B,C,D)$ where the input is that of the first subsystem and the output is that of the hth subsystem. The complete set of unobservable and uncontrollable modes of $S(A,B,C,D)$ is the set of zeros of the polynomial $p_d(s)$ where

$$p_d(s) \quad = \quad p_x(s)p_{d_1}(s)p_{d_2}(s)\ldots p_{d_h}(s) \qquad (4.2.9)$$

The open-loop characteristic polynomial can therefore be rewritten as

$$\text{OLCP}(s) = p_G(s)p_d(s) \qquad (4.2.10)$$

or

$$\text{OLCP}(s) = p_G(s)p_x(s)p_{d_1}(s)p_{d_2}(s)\ldots p_{d_h}(s) \qquad (4.2.11)$$

and if we combine expression (4.2.10) with the fundamental relationship (4.2.1) we obtain

$$\text{CLCP}(s) \quad = \quad p_d(s)p_G(s)\frac{\det F(s)}{\det F(\infty)} \qquad (4.2.12)$$

Now from the Smith-McMillan canonical form for $F(s)$ we have that

$$\det F(s) \quad = \quad \beta.\,\frac{z_F(s)}{p_F(s)} \qquad (4.2.13)$$

where β is a scalar quantity independent of s; and from the structure of $F(s)$, equation (3.1.2), it is clear that the monic polynomials $z_F(s)$ and $p_F(s)$ will be of the same order and hence that

$$\det F(\infty) \quad = \quad \beta \qquad (4.2.14)$$

Therefore combining equations (4.2.12-14) the closed-loop characteristic polynomial is given by

$$\text{CLCP (s)} = p_d(s)p_G(s)\ \frac{z_F(s)}{p_F(s)} \qquad (4.2.15)$$

If we now substitute from equations (4.2.4) and (3.3.35) this expression can be rewritten as follows

$$\text{CLCP(s)} = p_d(s)e_G(s)\ \frac{\prod\limits_{i=1}^{\ell} b'_{io}(s)\ \prod\limits_{i=1}^{\ell} d'_{it_i}(s)}{\prod\limits_{i=1}^{\ell} d'_{io}(s)} \qquad (4.2.16)$$

But using the eigenvalue shift theorem [8], on equation (3.1.2), we have that the characteristic functions $\{f_i(s): \ i=1,2,\ldots,\ell\}$ of F(s) are related to the characteristic gain functions $\{g_i(s): \ i=1,2,\ldots,\ell\}$ by

$$f_i(s) = 1+kg_i(s), \qquad i=1,2,\ldots,\ell \qquad (4.2.17)$$

and hence that the pole polynomials for both sets of algebraic functions are identical, that is

$$\prod\limits_{i=1}^{\ell} d'_{io}(s) = \prod\limits_{i=1}^{\ell} b'_{io}(s) \qquad (4.2.18)$$

and therefore from (4.2.16) we have

$$\boxed{\text{CLCP(s)} = p_d(s)e_G(s)\ \prod\limits_{i=1}^{\ell} d'_{it_i}(s)} \qquad (4.2.19)$$

Note that the polynomial $\prod\limits_{i=1}^{\ell} d'_{it_i}(s)$ is dependent on the gain control variable k.

The relationship (4.2.19) implies [see section 2.3] that the following conditions are necessary and sufficient for closed-loop stability:

(a) $e_G(s) = 0$ has only left half-plane roots;

(b) $\prod\limits_{i=1}^{\ell} d'_{it_i}(s) = 0$ has only left half-plane roots; and

(c) $p_d(s) = 0$ has only left half-plane roots.

The next step in the proof is to show how condition
(b) can be replaced by an encirclement condition similar
to that in the classical Nyquist criterion.

For the set of irreducible characteristic equations
associated with the return-difference operator F(s) there
is a corresponding set of Riemann surfaces on which the
appropriate characteristic algebraic functions $\{f_i(s): i=1,2,\ldots,\ell\}$
become single-valued, and mappings from these surfaces on
to a corresponding f_i - plane are one-to-one and continuous.
Let us consider the jth equation of the set defined by
equations (4.2.3). The degree of the equation is t_j and
therefore the corresponding Riemann surface $\mathbb{R}_{f_j}$ is formed
by piecing together t_j copies of the complex s-plane, $\mathbb{C}$.
Suppose now that a Nyquist D-contour, as shown in figure 16,
is drawn on each of the t_j copies of $\mathbb{C}$ before they are
pieced together to form $\mathbb{R}_{f_j}$. Then when the surface is
formed the set of Nyquist D-contours combine to form a set
of closed Jordan contours [9] enclosing right half-plane
regions of $\mathbb{R}_{f_j}$. The extended Principle of the Argument
[appendix 5] can then be applied to each right half-plane
region on $\mathbb{R}_{f_j}$. Therefore for a particular right half-
plane region, not necessarily simply connected but with a
boundary made up from Nyquist D-contours, we have that the
difference between the number of zeros and poles of the
algebraic function $f_j(s)$ in the region, is equal to the
number of clockwise encirclements of the origin in $\mathbb{C}$ (the
complex f-plane) by the image of the boundary curves, under
$f_j(s)$ for that particular region. If we therefore consider

all the right half-plane regions on $\mathbb{R}_{f_j}$ and apply the
extended Argument Principle to each, we have that

$$N(f_j \ , \ 0) \quad = \quad Z_{f_j} \quad - \quad P_{f_j} \qquad\qquad (4.2.20)$$

where:

(i) $N(f_j,0)$ is the net sum of clockwise encirclements
of the origin in $\mathbb{C}$ by the set of image curves, under $f_j(s)$,
of the set of curves formed on $\mathbb{R}_{f_j}$ by piecing together
the appropriate set of Nyquist D-contours when forming $\mathbb{R}_{f_j}$;

(ii) Z_{f_j} is the number of right half-plane zeros of $f_j(s)$; and

(iii) P_{f_j} is the number of right half-plane poles of $f_j(s)$.

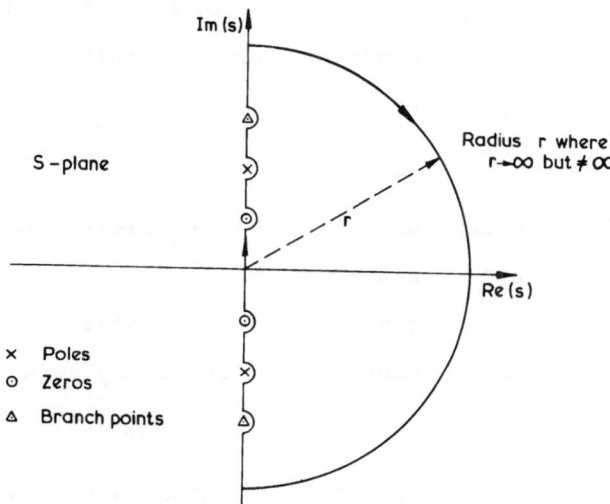

Figure 16. Theoretical Nyquist D-contour

Note that the indentations on the D-contour are necessary only from a theoretical viewpoint in the development of the Extended Principle of the Argument [appendix 5] and in practice the D-contour is taken as the imaginary axis.

Condition (b) for closed-loop stability is equivalent to saying that there are no zeros of

$$\{f_i(s): \quad i=1,2,\ldots,\ell\} \qquad (4.2.21)$$

in the right half-plane or on the imaginary axis, and can therefore be replaced by

$$Z_{f_i} = 0 \quad i=1,2,\ldots,\ell$$

or $\qquad\qquad\qquad\qquad\qquad\qquad\qquad\qquad (4.2.22)$

$$N(f_i,0) = -P_{f_i} \qquad i=1,2,\ldots,\ell$$

plus the condition that the algebraic functions (4.2.21) have no zeros on the imaginary axis. Conditions (4.2.22) imply and are implied by

$$\sum_{i=1}^{\ell} N(f_i,0) = -\sum_{i=1}^{\ell} P_{f_i} \qquad (4.2.23)$$

so that the necessary and sufficient conditions for closed-loop stability can be rewritten as

(a') $e_G(s)=0$ has no right half-plane roots;

(b') $e_{\ell G}(s)=0$ has no roots on the imaginary axis;

(c') $\sum_{i=1}^{\ell} N(f_i,0) = -\sum_{i=1}^{\ell} P_{f_i}$;

(d') $\{f_i(s): i=1,2,\ldots,\ell\}$ have no zeros on the imaginary axis; and

(e') $p_d(s)$ has only left half-plane zeros.

Now, as shown by equations (3.3.35) and (4.2.18), $e(s)$ together with the pole polynomials for the set of characteristic functions $\{f_i(s): i=1,2,\ldots,\ell\}$ make up the pole polynomial for $G(s)$. This leads us to consider combining conditions (a') and (c') into the single equivalent condition

$$\sum_{i=1}^{\ell} N(f_i, 0) = -P_G \tag{4.2.24}$$

where P_G is the number of right half-plane poles of $G(s)$.
The necessity and sufficiency of condition (4.2.24) for
closed-loop stability when conditions (b'),(d') and (e')
are satisfied is proved as follows.

From equations (3.3.35) and (4.2.18) we have that

$$P_G = e + \sum_{i=1}^{\ell} P_{f_i} \tag{4.2.25}$$

where e is the number of right half-plane zeros of $e_G(s)$,
and we also know that

$$\sum_{i=1}^{\ell} N(f_i, 0) = \sum_{i=1}^{\ell} Z_{f_i} - \sum_{i=1}^{\ell} P_{f_i} \tag{4.2.26}$$

so that combining these two expressions we have

$$\sum_{i=1}^{\ell} N(f_i, 0) = \sum_{i=1}^{\ell} Z_{f_i} + e - P_G \tag{4.2.27}$$

where $\sum_{i=1}^{\ell} Z_{f_i}$, e and P_G are all positive integers, or zero.

To establish the necessity of condition (4.2.24)
suppose that

$$\sum_{i=1}^{\ell} N(f_i, 0) \neq -P_G$$

then from equation (4.2.27) this implies that

$$\sum_{i=1}^{\ell} Z_{f_i} \neq 0 \quad \text{and/or} \quad e \neq 0$$

and hence we conclude that condition (4.2.24) is necessary
for closed-loop stability.

For sufficiency suppose that

$$\sum_{i=1}^{\ell} N(f_i, 0) = -P_G$$

then from equation (4.2.27) we must have that

$$\sum_{i=1}^{\ell} Z_{f_i} = e = 0$$

and hence the system is closed-loop stable.   Thus the sufficiency of condition (4.2.24) is established.

We have therefore shown that the following conditions are necessary and sufficient for closed-loop stability:

(a") $\sum_{i=1}^{\ell} N(f_i, 0) = - P_G$;

(b") $\{f_i(s): i=1,2,\ldots,\ell\}$ have no zeros on the imaginary axis;

(c") $e_G(s)$ has no zeros on the imaginary axis; and

(d") $P_d(s)$ has only left half-plane zeros.

From equation (4.2.17) the image curve sets in $\mathbb{C}$ of the Nyquist D-contour set mapped under $f_i(s)$ and $kg_i(s)$ are simply related by a unit shift in $\mathbb{C}$.   The stability condition (a") can consequently be replaced by

$$\sum_{i=1}^{\ell} N(kg_i, -1) = -P_G \qquad (4.2.28)$$

where $N(kg_i, -1)$ is the net sum of clockwise encirclements of the point $(-1+j0)$ in $\mathbb{C}$ by the characteristic gain loci scaled by k.   As in the classical Nyquist criterion the scaling by k is avoided by counting the number of encirclements that the characteristic gain loci make of the critical point $(-\frac{1}{k} + j0)$, and hence replacing condition (4.2.28) by

$$\sum_{i=1}^{\ell} N(g_i, -\frac{1}{k}) = -P_G \qquad (4.2.29)$$

From equations, (4.2.17) we also have that condition (b") is equivalent to

$$g_i(jw) \neq -\frac{1}{k}, \quad i=1,2,\ldots,\ell \qquad (4.2.30)$$

and in practice this corresponds to the characteristic gain loci not passing through the critical point $(-\frac{1}{k}+j0)$.

Condition (c") can also be replaced by a more practical

test.   The pole polynomial for G(s) is given, equation
(3.3.35), as

$$P_G(s) = e_G(s) \prod_{i=1}^{\ell} b'_{io}(s) \qquad\qquad (4.2.31)$$

where $b'_{jo}(s)$ is the pole polymomial for the jth characteristic
gain function $g_j(s)$.   Therefore it is clear that $e_G(s)$
will have no zeros on the imaginary axis if, and only if,
the number of infinite branches of the characteristic gain
loci is equal to the number of poles of G(s) on the imaginary
axis.   Note that for a branch of the loci going off to
infinity there will also be a branch returning from infinity
and care should be taken not to count this as two infinite
branches.

The necessary and sufficient conditions for closed-
loop stability can therefore be reduced to:

(1) $\sum_{i=1}^{\ell} N(g_i, -\frac{1}{k}) = - P_G;$

(2)   the characteristic gain loci do not pass through the
critical point $(-\frac{1}{k}+j0)$;

(3)   the number of branches of the characteristic gain loci
passing through infinity is equal to the number of poles of
G(s) on the imaginary axis;   and

(4) $p_d(s)$ has only left half-plane zeros.

This completes the proof of statement 1 of the generalized
Nyquist stability criterion.   ■

If the subsystems are completely characterized by
their transfer function matrices then their state-space
descriptions will each be observable and controllable so
that

$$P_{d_1}(s) = P_{d_2}(s) = \ldots = P_{d_h}(s) = 1 \quad (4.2.32)$$

and hence

$$P_d(s) = P_x(s) \quad (4.2.32)$$

Condition (4) of statement 1 of the Nyquist criterion is
then equivalent to $p_x(s)$ having only left half-plane zeros.
This situation also arises if the subsystems are characterized
by their state-space descriptions and we have the additional
information that each subsystem has no unobservable or
uncontrollable modes in the right half-plane or on the
imaginary axis.   In these situations conditions (1) and
(4) can be combined to give a criterion which needs no
information about decoupling zeros.

To show this we will first of all prove that when

(i) $\quad P_d(s) = P_x(s) \quad\quad\quad (4.2.34)$

or

(ii) $\quad P_d(s) = P_x(s)P_{d_1}(s)P_{d_2}(s) \ldots P_{d_h}(s)$
$\quad\quad\quad\quad\quad\quad\quad\quad\quad\quad\quad (4.2.35)$

where the zeros of $\{p_{d_i}(s): i=1,2,\ldots,h\}$ are
all in the left half-plane,

conditions (a'), (c') and (e') are equivalent to

$$\sum_{i=1}^{\ell} N(f_i,0) = -\sum_{i=1}^{h} P_{G_i} \quad (4.2.36)$$

where P_{G_i} is the number of right half-plane poles of $G_i(s)$.
The necessity and sufficiency of condition (4.2.36) for

closed-loop stability when conditions (b') and (d') are satisfied is proved as follows.

From equations (4.2.9),(4.2.18) and (4.2.31) we have that

$$\sum_{i=1}^{h} P_{G_i} = e + \sum_{i=1}^{\ell} P_{f_i} + P_x \qquad (4.2.37)$$

where P_x is the number of right half-plane zeros of $P_x(s)$, and combining this with equation (4.2.26) we have that

$$\sum_{i=1}^{\ell} N(f_i,O) = \sum_{i=1}^{\ell} Z_{f_i} + e + P_x - \sum_{i=1}^{h} P_{G_i} \qquad (4.2.38)$$

Note also that

$$P_d(s) = P_x(s)P_{d_1}(s) \ldots P_{d_h}(s)$$

so that if P_d denotes the number of right half-plane zeros of $P_d(s)$ we have that

$$P_d = P_x \qquad (4.2.39)$$

and equation (4.2.38) becomes

$$\sum_{i=1}^{\ell} N(f_i,O) = \sum_{i=1}^{\ell} Z_{f_i} + e + P_d - \sum_{i=1}^{h} P_{G_i} \qquad (4.2.40)$$

To establish the necessity of condition (4.2.36) suppose that

$$\sum_{i=1}^{\ell} N(f_i,O) \neq - \sum_{i=1}^{h} P_{G_i}$$

then from equation (4.2.40) this implies that

$$\sum_{i=1}^{\ell} Z_{f_i} \neq O, \text{ or } e \neq O, \text{ or } P_d \neq O$$

or any combination of these and thus that the system will be closed-loop unstable. Hence we conclude that condition (4.2.36) is necessary for closed-loop stability.

For sufficiency suppose that

$$\sum_{i=1}^{\ell} N(f_i,O) = - \sum_{i=1}^{h} P_{G_i}$$

then from equation (4.2.40) we must have that

$$\sum_{i=1}^{\ell} Z_{f_i} = e = P_d = 0$$

and hence the system is closed-loop stable.   Thus the
sufficiency of condition (4.2.36) is established.

   We have therefore shown that when either condition
(4.2.34) or condition (4.2.35) is satisfied the following
are necessary and sufficient for closed-loop stability.

(a''') $\sum_{i=1}^{\ell} N(f_i, 0) = - \sum_{i=1}^{h} P_G$;

(b''') $\{f_i(s): i=1,2,\ldots,\ell\}$ have no zeros on the imaginary
axis;   and

(c''') $e_G(s)$ has no zeros on the imaginary axis.

But we have already shown, in proving statement 1 of the
stability criterion, that

$$\sum_{i=1}^{\ell} N(f_i, 0) \equiv \sum_{i=1}^{\ell} N(g_i, -\frac{1}{k})   ;$$

that condition (b''')   is equivalent to the characteristic
gain loci not passing through the critical point $(-\frac{1}{k}+j0)$;
and that condition (c''')   is equivalent to the number of
infinite branches of the characteristic gain loci being
equal to the number of poles of G(s) on the imaginary axis.
This therefore completes the proof of statment 2 of the
generalized Nyquist stability criterion.   ∎

4.3   Example

   To illustrate the stability criterion consider the
example already used in sub-sections 3.3-5 and 3.4-2, where
the general feedback configuration is characterized by its
open-loop gain matrix

$$G(s) = \frac{1}{1.25(s+1)(s+2)} \begin{bmatrix} s-1 & s \\ -6 & s-2 \end{bmatrix}$$

By definition the system has no unobservable or uncontrollable modes and by virtue of the problem statement we can consider the configuration as consisting of just one subsystem $G_1(s)=G(s)$. Therefore either statement 1 or statement 2 of the stability criterion is directly applicable.

The pole polynomial for $G(s)$ is

$$P_G(s) = (s+1)(s+2)$$

which has no zeros in the right half-plane or on the imaginary axis and therefore closed-loop stability is ensured if the net sum of anti-clockwise encirclements of the critical point $(-\frac{1}{k}+j0)$ by the characteristic gain loci is zero, and if the characteristic gain loci do not pass through the critical point and have no infinite branches.   The characteristic gain loci are shown in figure 17 from which the following. stability conditions are obtained.

(i)   For $-\infty \leqslant -\frac{1}{k} < -0.8$ there are no encirclements of, or passage through, the critical point and thus the closed-loop system is stable for $0 \leqslant k<1.25$.

(ii)   For $-\frac{1}{k} = -0.8$ the characteristic gain loci pass through the critical point and therefore the system is closed-loop unstable for $k=1.25$

(iii) For $-0.8<-\frac{1}{k}<-0.4$ there is one clockwise encirclement of the critical point and therefore the system is closed-loop unstable for $1.25<k<2.5$

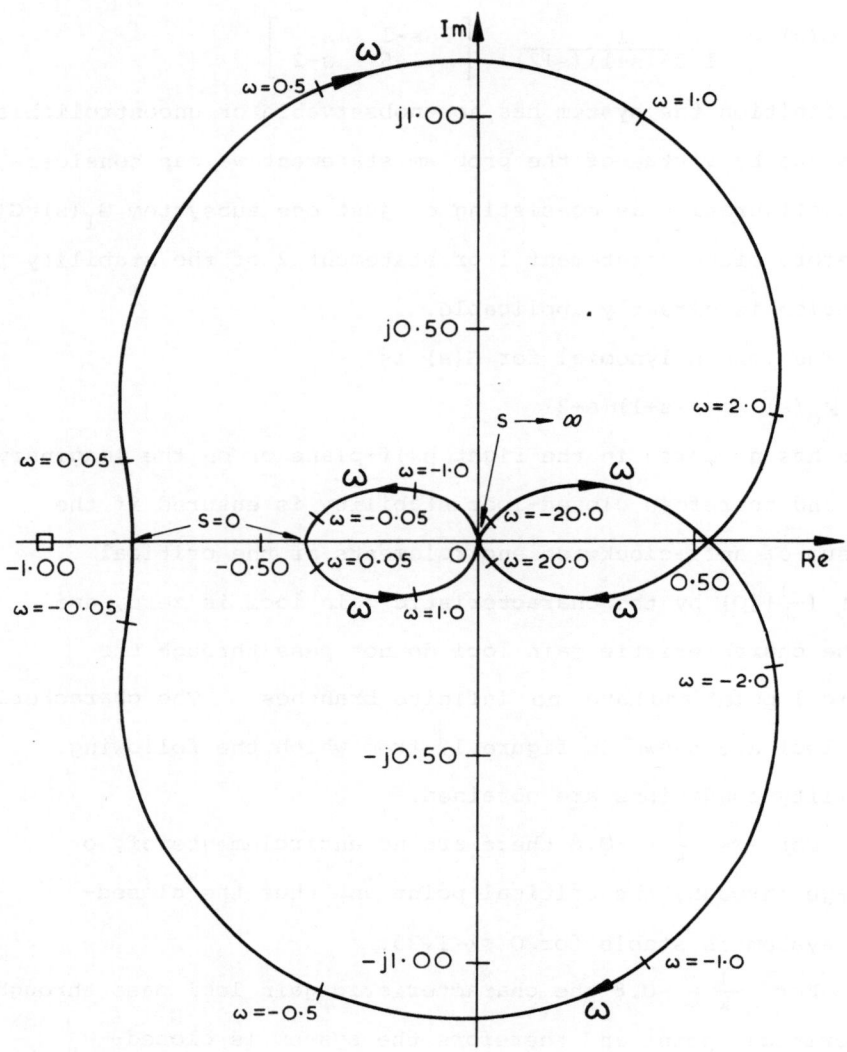

Figure 17. Characteristic gain loci

(iv)   For $-\frac{1}{k} = -0.4$ the characteristic gain loci pass through the critical point and therefore the system is closed-loop unstable for k=2.5.

(v)   For $-0.4 < -\frac{1}{k} < 0$ there are no encirclements of, or passage through, the critical point and thus the closed-loop system is stable for 2.5<k<∞.

(vi)   For $-\frac{1}{k} = 0$ the characteristic gain loci pass through the critical point and therefore the system is closed-loop unstable for k=∞.

(vII)   For $0 < -\frac{1}{k} < 0.533$ there are two clockwise encirclements of the critical point and therefore the system is closed-loop unstable for -∞<k<-1.875.

(viii) For $-\frac{1}{k} = 0.533$ the characteristic gain loci pass through the critical point and therefore the system is closed-loop unstable for k=-1.875.

(ix)   For $0.533 < -\frac{1}{k} \leqslant \infty$ there are no encirclements of, or passage through, the critical point and thus the closed-loop system is stable for -1.875<k≤0.

Note that the conditions where k is negative correspond to positive feedback.

References

[1] H. Nyquist, "The Regeneration Theory", Bell System Tech. J., 11, 126-147, 1932.

[2] A.G.J. MacFarlane, "Return-difference and return-ratio matrices and their use in analysis and design of multivariable feedback control systems", Proc. IEE, 117, 2037-2049, 1970.

[3] A.G.J. MacFarlane and J.J. Belletrutti, "The characteristic Locus Design Method", Automatica, 9, 575-588, 1973.

[4] J.F. Barman and J. Katzenelson, Memorandum ERL-383, Electronics Research Laboratory, College of Engineering, Univ. of California, Berkeley, 1973.

[5] J.F. Barman and J. Katzenelson, "A generalized Nyquist-type stability criterion for multivariable feedback systems", Int. J. Control, 20, 593-622, 1974.

[6] A.G.J. MacFarlane and N. Karcanias, "Poles and zeros of linear multivariable systems: a survey of the algebraic, geometric and complex variable theory", Int. J. Control, 24, 33-74, 1976.

[7] C.A. Desoer and W.S. Chan, "The Feedback Interconnection of Lumped Linear Time-invariant Systems", J. Franklin Inst., 300, 335-351, 1975.

[8] A.G.J. MacFarlane, "Dynamical System Models", Harrap, London, 1970.

[9] E. Hille, "Analytic Function Theory", Vol. 1, Ginn and Co., U.S.A., 1959.

5.   A generalized inverse Nyquist stability criterion

In this chapter a generalization of the inverse Nyquist stability criterion [1] for single-input single-output feedback systems is developed for the general feedback configuration which is complementary to the exposition of the generalized Nyquist stability criterion presented in chapter 4.   The development is based on an association of the open-loop gain matrix G(s) with a set of inverse characteristic gain functions, and a corresponding set of inverse Nyquist diagrams which will be termed the inverse characteristic gain loci.

5.1   Inverse characteristic gain functions

A   main feature of the generalization of Nyquist's stability criterion, given in chapter 4, is the association of a set of algebraic functions - the characteristic gain functions - with a square transfer function matrix G(s) by means of the characteristic equation

$$\Delta(g,s) = \det\left[gI_m - G(s) \right] = 0 \qquad (5.1.1)$$

If G(s) has normal rank m then its inverse function $G(s)^{-1}$ exists, and has a corresponding characteristic equation given by

$$\Lambda(\overset{*}{g},s) = \det\left[\overset{*}{g}I_m - G(s)^{-1}\right] = 0 \qquad (5.1.2)$$

If $\Lambda(\overset{*}{g},s)$ is regarded as a polynomial in $\overset{*}{g}$ with coefficients which are rational functions of s then equation (5.1.2) defines a set of algebraic functions $\{\overset{*}{g}_i(s): i=1,2,\ldots,\ell\}$ which will be called the <u>inverse characteristic gain functions</u>. If $\Lambda(\overset{*}{g},s)$ is also irreducible over the field of rational functions in s, then the inverse characteristic gain function $\overset{*}{g}(s)$, like g(s), has as its domain an appropriate Riemann surface formed out of m copies of the complex s-plane suitably joined together. In fact, because the eigenvalues of a matrix are reciprocal to the eigenvalues of the matrix inverse

$$\overset{*}{g}(s) = \frac{1}{g(s)} \qquad\qquad (5.1.3)$$

and therefore $\overset{*}{g}(s)$ and g(s) have the same branch points, and hence the same Riemann surface domains.

The inverse characteristic gain function is the foundation on which the generalized inverse Nyquist stability criterion is based.

5.2   Pole-zero relationships

In this section a number of pole-zero relationships are derived which will be used in the proof of the generalized inverse Nyquist stability criterion, section 5.4.

Because the eigenvalues of a matrix are the reciprocal of the eigenvalues of the matrix inverse the poles of the set of inverse characteristic gain functions are simply the zeros of the characteristic gain function set, and vice versa. The pole and zero polynomials of $\overset{*}{g}_i(s)$, denoted by $p_g^*(s)$ and $z_g^*(s)$, are therefore expressible as

$$p_{g_i}^* (s) = z_{g_i} (s) \qquad\qquad (5.1.4)$$

and

$$z_{g_i}^* (s) = p_{g_i} (s) \qquad\qquad (5.1.5)$$
$$i=1,2,\ldots,\ell$$

From sub-section 3.3-2 it is clear that for an open-loop gain matrix G(s) there exists a canonical form, the

Smith-McMillan form $[2]$, such that

$$G(s) = H_G(s)M_G(s)J_G(s) \qquad (5.1.6)$$

where $H_G(s)$ and $J_G(s)$ are both mxm unimodular matrices and the Smith-McMillan form $M_G(s)$ is given by

$$M_G(s) = \text{diag}\left[\frac{\varepsilon_1(s)}{\psi_1(s)}, \frac{\varepsilon_2(s)}{\psi_2(s)}, \ldots \frac{\varepsilon_m(s)}{\psi_m(s)} \right] \qquad (5.1.7)$$

Consequently the poles and zeros of G(s) are defined as

$$P_G(s) = \prod_{i=1}^{m} \psi_i(s) \qquad (5.1.8)$$

and

$$z_G(s) = \prod_{i=1}^{m} \varepsilon_i(s) \qquad (5.1.9)$$

Now if G(s) is of normal rank m equation (5.1.6) can be inverted to give

$$G(s)^{-1} = J_G(s)^{-1}M_G(s)^{-1}H_G(s)^{-1} \qquad (5.1.10)$$

which using the elementary transformation matrix E, of order m, and given by

$$E = \begin{bmatrix} 0 & 0 & \cdots & 0 & 1 \\ 0 & 0 & \cdots & 1 & 0 \\ \vdots & \vdots & & \vdots & \vdots \\ 0 & 1 & \cdots & 0 & 0 \\ 1 & 0 & \cdots & 0 & 0 \end{bmatrix} = E^{-1} \qquad (5.1.11)$$

can be rewritten as

$$G(s)^{-1} = J_G(s)^{-1}EEM_G(s)^{-1}EEH_G(s)^{-1} \qquad (5.1.12)$$

or

$$G(s)^{-1} = H_G^*(s)M_G^*(s)J_G^*(s) \qquad (5.1.13)$$

where

$$H_G^*(s) = J_G(s)^{-1}E \qquad (5.1.14)$$

$$J_G^*(s) = EH_G(s)^{-1} \qquad (5.1.15)$$

and

$$M_G^*(s) = EM_G(s)^{-1}E$$

$$= \text{diag}\left[\frac{\psi_m(s)}{\varepsilon_m(s)} , \frac{\psi_{m-1}(s)}{\varepsilon_{m-1}(s)} , \dots, \frac{\psi_1(s)}{\varepsilon_1(s)}\right] \qquad (5.1.16)$$

which is the Smith-McMillan form for $G(s)^{-1}$. The pole-zero definitions for a transfer function matrix [sub-section 3.3-2] can now be applied to $G^{-1}(s)$ with the result that

$$p_G^*(s) = \prod_{i=1}^{m} \varepsilon_i(s) = z_G(s) \qquad (5.1.17)$$

and

$$z_G^*(s) = \prod_{i=1}^{m} \psi_i(s) = p_G(s) \qquad (5.1.18)$$

where $p_G^*(s)$ and $z_G^*(s)$ denote the pole and zero polynomials of $G(s)^{-1}$ respectively.

In sub-section 3.3-3 it is shown that the pole and zero polymomials of $G(s)$ are related to the poles and zeros of the characteristic gain function set $\{g_i(s):i=1,2,\dots,\ell\}$ by

$$p_G(s) = e_G(s) \prod_{i=1}^{\ell} b_{io}'(s) = e_G(s) \prod_{i=1}^{\ell} p_{g_i}(s) \qquad (5.1.19)$$

and

$$z_G(s) = e_G(s) \prod_{i=1}^{\ell} b_{it_i}'(s) = e_G(s) \prod_{i=1}^{\ell} z_{g_i}(s) \qquad (5.1.20)$$

If the pole and zero polynomials of the complete characteristic gain function set are denoted by $p_g(s)$ and $z_g(s)$ respectively, and similarly for the inverse characteristic gain function set using $p_g^*(s)$ and $z_g^*(s)$, then relationships (5.1.19) and

(5.1.20) can be combined with relationships (5.1.17) and

(5.1.18) to give

$$P_G(s) = z_G^*(s) = e_G(s) p_g(s) = e_G(s) z_g^*(s) \qquad (5.1.21)$$

and

$$z_G(s) = p_G^*(s) = e_G(s) z_g(s) = e_G(s) p_g^*(s) \qquad (5.1.22)$$

These pole-zero relationships will be used in the proof of

the inverse Nyquist stability criterion presented later.

5.3   Inverse characteristic gain loci - generalized inverse Nyquist diagrams

The inverse characteristic gain loci are the loci in

the complex plane traced out by the reciprocal of each of

the eigenvalues of the open-loop gain matrix G(s) as s

traverses the Nyquist D-contour in the standard (clockwise)

direction.   The algorithm for computing the loci follows

that given for the characteristic gain loci, in sub-section

3.4-1, with the modification that at step (iv) the reciprocal

of the numbers $\bar{g}_i(jwa)$ are plotted in the complex plane.

We are now in a position to state and prove the

generalized inverse Nyquist stability criterion.

5.4   Generalized inverse Nyquist stability criterion

If the subsystems of the general feedback configuration

of figure 3 are each characterized by a state-space model

then the following statement of the criterion is applicable.

Statement 1.   The general feedback configuration is closed-

loop stable if and only if:

(1a)   the net sum of anti-clockwise encirclements, of the

critical point $(-k+jo)$, by the set of inverse characteristic gain loci, minus the net sum of anti-clockwise encirclements, of the origin, by the set of inverse characteristic gain loci, is equal to the number of right half-plane poles of G(s);

(2)   the inverse characteristic gain loci do not pass through the critical point (-k+jo); and

(3a) the number of branches of the inverse characteristic gain loci that pass through the origin is equal to the number of poles of G(s) on the imaginary axis; and

(4)   the eigenvalues of the A- matrix, of the open-loop system S(A,B,C,D), which correspond to modes of the system which are unobservable and/or uncontrollable from the point of view of considering the input as that of the first subsystem and the output as that of the hth subsystem, are all in the left half-plane.

Alternatively conditions (1a) and (3a) may be replaced by:

(1b) the net sum of anti-clockwise encirclements of the critical point (-k+jo) by the set of inverse characteristic gain loci is equal to the number of right half-plane zeros of G(s);

(3b) the number of branches of the inverse characteristic gain loci passing through infinity is equal to the number of zeros of G(s) on the imaginary axis.

If the subsystems are completely characterized by their transfer function matrices, or if it is known that for each subsystem there are no unobservable and/or uncontrollable modes in the right half-plane including the imaginary axis, then the following statement of the criterion applies.

Statement 2   The general feedback configuration is closed-loop stable if and only if:

(1a)   the net sum of anti-clockwise encirclements of the critical point (-k+jo), by the set of inverse characteristic gain loci, minus the net sum of anti-clockwise encirclements, of the origin, by the set of inverse characteristic gain loci, is equal to the total number of right half-plane poles of $G_1(s), G_2(s), \ldots,$ and $G_h(s);$

(2)   the inverse characteristic gain loci do not pass through the critical point (-k+jo);   and

(3a)   the number of branches of the inverse characteristic gain loci that pass through the origin is equal to the number of poles of G(s) on the imaginary axis.

Alternatively condition (3a) may be replaced by condition (3b), as in statement 1 of the criterion, and if the subsystems are square, i.e. have the same number of outputs as inputs, then condition (1a) may be replaced by:

(1b)   the net sum of anti-clockwise encirclements of the critical point (-k+jo) by the set of inverse characteristic gain loci is equal to the total number of right half-plane zeros of $G_1(s), G_2(s), \ldots,$ and $G_h(s)$.

Note that if condition (2) and/or condition (3a)/(3b) do not hold then the closed-loop system has one or more poles on the imaginary axis and is therefore not input-output stable although the equilibrium state at the origin may be stable.

For strictly proper systems, that is, ones in which the system D-matrix is zero, the inverse characteristic gain loci

approach infinity as s approaches infinity, and s=∞ is
a pole of the inverse function.   Therefore, in practice, it
is necessary to traverse the whole of the D-contour, not
just the imaginary axis, in order to obtain closed curves.
In this way the net sum of encirclements of the critical
point and the origin by the inverse characteristic gain loci
can be obtained.   As an example, the inverse characteristic
gain loci for

$$G(s) \; = \; \frac{1}{(s+1)^3} \qquad\qquad (5.4.1)$$

are shown in figure 18.

The traversal of the D-contour off the imaginary axis,
however, is not necessary if we use conditions (1a), rather
than (1b), in both statements of the stability criterion.
This is because the extra loci, corresponding to the circular
part of the D-contour, encircle the origin and the critical
point the same number of times, and thereby cancel each other.
A useful rule, when using the (1a) conditions, is therefore
to join up the corresponding loose ends of the loci via a
large semi circle, and to forget any extra encirclements that
may actually exist.   Whether the ends are joined via the
right half-plane or left half-plane is fixed by the fact that
the mapping under $\overset{*}{g}(s)$, from the set of Nyquist curves formed
on the Riemann surface of $\overset{*}{g}(s)$ (by piecing together an appropriate
set of Nyquist D-contours) to the $\overset{*}{g}$-plane, is conformal.   This
means that if we imagine two people, X and Y, X walking along
the domain curve, and Y walking along the corresponding image
curve, such that at each step X defines the corresponding
image point, then if X turns to his right, Y will turn to his
right.

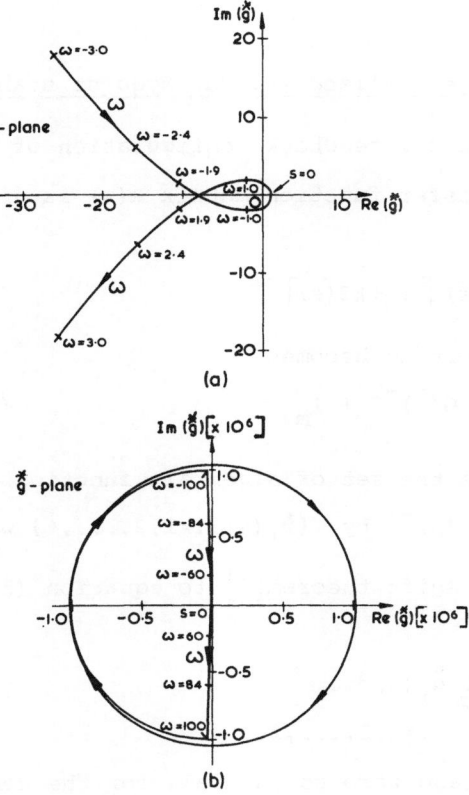

Figure 18. Inverse characteristic gain
loci for $G(s) = 1 / (s+1)^3$
(a) small area around origin
(b) large area

5.5   Proof of generalized inverse Nyquist stability criterion

For the general feedback configuration of figure 3 the closed-loop transfer function matrix R(s) is given by the relationship

$$R(s) = kG(s) \left[I_m + kG(s) \right]^{-1} \qquad (5.5.1)$$

which after inverting becomes

$$R(s)^{-1} = \frac{1}{k} G(s)^{-1} + I_m \qquad (5.5.2)$$

If we now denote the set of algebraic functions defining the eigenvalues of $R(s)^{-1}$ by $\{\overset{*}{r}_i(s):i=1,2,\ldots,\ell\}$ we can apply the eigenvalue shift theorem[3] to equation (5.5.2) with the result that

$$\overset{*}{r}_i(s) = \frac{1}{k} \overset{*}{g}_i(s) + 1 \qquad (5.5.3)$$
$$i=1,2,\ldots,\ell$$

Then if the pole and zero polynomials for the complete set of algebraic functions $\{\overset{*}{r}_i(s):i=1,2,\ldots,\ell\}$ are denoted by the monic polynomials $p_r^*(s)$ and $z_r^*(s)$ respectively, we have that

$$p_r^*(s) = p_g^*(s) \qquad (5.5.4)$$

Now post-multiplying equation (5.5.2) by kG(s) we have

$$kR(s)^{-1}G(s) = I_m + kG(s)$$
$$= F(s) \qquad (5.5.5)$$

the return difference matrix [4], and taking determinants of this we obtain

$$\det F(s) = k^m \frac{\det\left[R(s)^{-1}\right]}{\det\left[G(s)^{-1}\right]}$$

$$= \gamma \frac{z_r^*(s)}{p_r^*(s)} \cdot \frac{p_g^*(s)}{z_g^*(s)} \qquad (5.5.6)$$

where γ is a scalar constant independent of s.   But

$$\det F(s) \;=\; \beta \; \frac{z_F(s)}{p_F(s)} \;=\; \beta \; \frac{z_f(s)}{p_f(s)} \qquad (5.5.7)$$

where $p_f(s)$ and $z_f(s)$ are respectively the monic pole and
zero polynomials for the set of algebraic functions $\{f_i(s):i=1,2,\ldots,\ell\}$
and therefore

$$\beta = \gamma \qquad (5.5.8)$$

and

$$z_f(s) \;=\; \frac{z_r^*(s)}{p_r^*(s)} \cdot \frac{p_g^*(s)}{z_g^*(s)} \cdot p_f(s) \qquad (5.5.9)$$

Now from equations (4.2.18) and (5.1.21)

$$p_f(s) \;=\; p_g(s) \;=\; z_g^*(s) \qquad (5.5.10)$$

and hence combining this with equations (5.5.4) and (5.5.9)
we obtain

$$\boxed{z_f(s) \;=\; z_r^*(s)} \qquad (5.5.11)$$

In chapter 4, equation (4.2.19), it was shown that the
closed-loop characteristic polynomial can be given by

$$CLCP(s) = p_d(s)e_G(s) \prod_{i=1}^{\ell} d_{it_i}'(s) \qquad (5.5.12)$$

or equivalently, using the notation presented in this chapter,

$$CLCP(s) \;=\; p_d(s)e_G(s)z_f(s) \qquad (5.5.13)$$

and therefore combining this with equation (5.5.11) we have

$$\boxed{CLCP(s) \;=\; p_d(s)e_G(s)z_r^*(s)} \qquad (5.5.14)$$

This implies that the following conditions are necessary and
sufficient for closed-loop stability:

(a) $e_G(s) = 0$ has only left half-plane roots;

(b) $z_r^*(s) = 0$ has only left half-plane roots;   and

(c) $p_d(s) = 0$ has only left half-plane roots.

Suppose now that a Nyquist D-contour, as shown in figure 16, is drawn on m copies of the complex plane before they are pieced together to form the Riemann surfaces $\{\mathbb{R}^*_{r_i} : i=1,2,\ldots,\ell\}$ on which the $\{r^*_i(s):i=1,2,\ldots,\ell\}$ are defined. Let us consider the jth surface $\mathbb{R}^*_{r_j}$ corresponding to $r^*_j(s)$. When the surface is formed the set of Nyquist D-contours combine to form a set of closed Jordan contours [5] enclosing right half-plane regions on $\mathbb{R}^*_{r_j}$. The extended Principle of the Argument [appendix 5] can be applied to each right half-plane region on $\mathbb{R}^*_{r_j}$. Therefore for a particular right half-plane region, not necessarily simply connected but with a boundary made up from Nyquist D-contours, we have that the difference between the number of zeros and poles of the algebraic function $r^*_j(s)$ in the region, is equal to the number of clockwise encirclements of the origin in $\mathbb{C}$ (the complex r^*-plane) by the image of the boundary curves, under $r^*_j(s)$, for that particular region.

If we therefore consider all the right half-plane regions and apply the extended Argument Principle to each, we have that

$$N(r^*_j, 0) = Z_{r^*_j} - P_{r^*_j} \qquad (5.5.15)$$

where:

(i) $N(r^*_j,0)$ is the net sum of clockwise encirclements of the origin in $\mathbb{C}$ by the set of image curves, under $r^*_j(s)$ of the set of curves formed on $\mathbb{R}^*_{r_j}$ by piecing together the appropriate set of Nyquist D-contours when forming $\mathbb{R}^*_{r_j}$;

(ii) $Z_{r^*_j}$ is the number of right half-plane zeros of $r^*_j(s)$;

and

(iii) $P^*_{r_j}$ is the number of right half-plane poles of $\overset{*}{r}_j(s)$.

If we now consider the corresponding inverse characteristic gain function $\overset{*}{g}_j(s)$ in the same way, we find that

$$N(\overset{*}{g}_j \, , \, 0) \; = \; Z^*_{g_j} \; - \; P^*_{g_j} \qquad\qquad (5.5.16)$$

where:

(i) $N(\overset{*}{g}_j,0)$ is the net sum of clockwise encirclements of the origin in $\mathbb{C}$ by the inverse characteristic gain loci corresponding to $\overset{*}{g}_j(s)$;

(ii) $Z^*_{g_j}$ is the number of right half-plane zeros of $\overset{*}{g}_j(s)$; and

(iii) $P^*_{g_j}$ is the number of right half-plane poles of $\overset{*}{g}_j(s)$.

Equations (5.5.15) and (5.5.16) can be combined to give

$$N(\overset{*}{r}_j,0)-N(\overset{*}{g}_j,0) = Z_{\overset{*}{r}_j} \; - \; P_{\overset{*}{r}_j} \; - \; Z_{\overset{*}{g}_j} \; + \; P_{\overset{*}{g}_j} \qquad (5.5.17)$$

which using equation (5.5.4) becomes

$$N(\overset{*}{r}_j,0) \; - \; N(\overset{*}{g}_j,0) \; = \; Z^*_{r_j} \; - \; Z^*_{g_j} \qquad\qquad (5.5.18)$$

and if we consider the complete set of algebraic functions $\{\overset{*}{r}_i(s):i=1,2,\ldots,\ell\}$ and the set $\{\overset{*}{g}_i(s):i=1,2,\ldots,\ell\}$

we have

$$\sum_{i=1}^{\ell} N(\overset{*}{r}_i,0) - \sum_{i=1}^{\ell} N(\overset{*}{g}_i,0) \; = \; \sum_{i=1}^{\ell} Z^*_{r_i} \; - \; \sum_{i=1}^{\ell} Z^*_{g_i} \qquad (5.5.19)$$

Now condition (b) for closed-loop stability is equivalent to saying that there are no zeros of $\{\overset{*}{r}_i(s):i=1,2,\ldots,\ell\}$

in the right half-plane or on the imaginary axis, and can therefore be replaced by

$$\sum_{i=1}^{\ell} N(\overset{*}{r}_i,0) \; - \; \sum_{i=1}^{\ell} N(\overset{*}{g}_i,0) \; = \; - \sum_{i=1}^{\ell} Z^*_{g_i} \qquad (5.5.20)$$

plus the condition that $\{\overset{*}{r}_i(s):i=1,2,\ldots,\ell\}$ have no zeros
on the imaginary axis.   The necessary and sufficient
conditions for closed-loop stability can therefore be
rewritten as

(a') $e_G(s) = 0$ has no right half-plane roots;

(b') $e_G(s) = 0$ has no roots on the imaginary axis;

(c') $\displaystyle\sum_{i=1}^{\ell} N(\overset{*}{r}_i,0) - \sum_{i=1}^{\ell} N(\overset{*}{g}_i,0) = - \sum_{i=1}^{\ell} z^*_{g_i}$;

(d') $\{\overset{*}{r}_i(s):i=1,2,\ldots,\ell\}$ have no zeros on the imaginary
axis;   and

(e') $p_d(s)$ has only left half-plane zeros.

The pole-zero relationship (5.1.21) now leads us to consider
combining conditions (a') and (c') into the single equivalent condition

$$\sum_{i=1}^{\ell} N(\overset{*}{r}_i,0) - \sum_{i=1}^{\ell} N(\overset{*}{g}_i,0) = -P_G \qquad (5.5.21)$$

where P_G is the number of right half-plane poles of $G(s)$.
The necessity and sufficiency of condition (5.5.21) for
closed-loop stability when conditions (b'), (d') and (e')
are satisfied is proved as follows

From relationship (5.1.21) we have that

$$P_G = e + \sum_{i=1}^{\ell} z^*_{g_i} \qquad (5.5.22)$$

where e is the number of right half-plane poles of $e_G(s)$, and
combining this with equation (5.5.19) gives

$$\sum_{i=1}^{\ell} N(\overset{*}{r}_i,0) - \sum_{i=1}^{\ell} N(\overset{*}{g}_i,0) = \sum_{i=1}^{\ell} z^*_{r_i} + e - P_G \qquad (5.5.23)$$

To establish the necessity of condition (5.5.21)
suppose that

$$\sum_{i=1}^{\ell} N(\overset{*}{r}_i,0) - \sum_{i=1}^{\ell} N(\overset{*}{g}_i,0) \neq -P_G$$

then from equation (5.5.23) this implies that

$$\sum_{i=1}^{\ell} z_{r_i}^* \neq 0 \qquad \text{and/or} \quad e \neq 0$$

and hence we conclude that condition (5.5.21) is necessary

for closed-loop stability.

For sufficiency suppose that

$$\sum_{i=1}^{\ell} N(\overset{*}{r}_i,0) - \sum_{i=1}^{\ell} N(\overset{*}{g}_i,0) = -P_G$$

then from equation (5.5.23) this implies that

$$\sum_{i=1}^{\ell} z_{\overset{*}{r}_i} = e = 0$$

and hence the system is closed-loop stable. Thus the sufficiency

of condition (5.5.21) is established.

We have therefore shown that the following conditions

are necessary and sufficient for closed-loop stability:

(a") $\sum_{i=1}^{\ell} N(\overset{*}{r}_i,0) - \sum_{i=1}^{\ell} N(\overset{*}{g}_i,0) = -P_G$;

(b") $\{\overset{*}{r}_i(s): i=1,2,\ldots,\ell\}$ have no zeros on the imaginary axis;

(c") $e_G(s)=0$ has no roots on the imaginary axis;   and

(d") $p_d(s)$ has only left half-plane zeros.

From equation (5.5.3) it is clear that

$$\sum_{i=1}^{\ell} N(\overset{*}{r}_i,0) = \sum_{i=1}^{\ell} N(\overset{*}{g}_i, -k) \qquad\qquad (5.5.24)$$

and consequently conditon (a") is equivalent to condition (1a)

in statement 1 of the stability criterion.

Also from equation (5.5.3) we have that $\{\overset{*}{r}_i(s): i=1,2,\ldots,\ell\}$

have zeros on the imaginary axis if, and only if the inverse

characteristic gain loci pass through the critical point

$(-k+j0)$;   hence condition (b") is equivalent to condition 2

in statement 1 of the stability criterion.

From the pole-zero relationships, (5.1.21) and (5.1.22),

condition (c") is clearly equivalent to condition (3a)
and also condition (3b) in either statement of the stability
criterion.

Therefore to complete the proof of statement 1 of the
stability criterion all that is left is to show that condition
(1b) is equivalent to condition (a"). To do this we will
show that when conditions (b"), (c") and (d") are satisfied,
the condition

$$\sum_{i=1}^{\ell} N(\overset{*}{r}_i, 0) = \sum_{i=1}^{\ell} N(\overset{*}{g}_i, -k) = -Z_G, \qquad (5.5.25)$$

where Z_G is the number of right half-plane zeros of $G(s)$,
is necessary and sufficient for closed-loop stability.

From relationship (5.1.22) we have that

$$Z_G = e + \sum_{i=1}^{\ell} P_{\overset{*}{g}_i} \qquad (5.5.26)$$

where $P_{\overset{*}{g}_j}$ is the number of right half-plane poles of $\overset{*}{g}_j(s)$,
but from equation (5.5.3) the poles of $\overset{*}{r}_i(s)$ and $\overset{*}{g}_i(s)$ are
the same, and hence

$$Z_G = e + \sum_{i=1}^{\ell} P_{\overset{*}{r}_i} \qquad (5.5.27)$$

If we now combine equation (5.5.27) with equation (5.5.15),
which is valid for $\{j=1,2,\ldots,\ell\}$, we have

$$\sum_{i=1}^{\ell} N(\overset{*}{r}_i, 0) = \sum_{i=1}^{\ell} Z_{\overset{*}{r}_i} + e - Z_G \qquad (5.5.28)$$

To establish the necessity of condition (5.5.25) suppose
that

$$\sum_{i=1}^{\ell} N(\overset{*}{r}_i, 0) \neq -Z_G$$

then from equation (5.5.28) this implies that

$$\sum_{i=1}^{\ell} Z_{\overset{*}{r}_i} \neq 0 \text{ and/or } e \neq 0$$

and hence we conclude that condition (5.5.28) is necessary
for closed-loop stability.

For sufficiency suppose that

$$\sum_{i=1}^{\ell} N(\overset{*}{r}_i, 0) = -z_G$$

then from equation (5.5.28) we have that

$$\sum_{i=1}^{\ell} z^*_{r_i} = e = 0$$

and hence that the system is closed-loop stable, providing
conditions (b"), (c") and (d") are satisfied. Thus the
sufficiency of condition (5.5.25) is established.

This completes the proof of statement 1 of the generalized
inverse Nyquist stability criterion. ∎

Statement 2 of the stability criterion applies to systems
[see chapter 4] in which either

(i) $p_d(s) = p_x(s)$ (5.5.29)

or

(ii) $p_d(s) = p_x(s) p_{d_1}(s) p_{d_2}(s) \ldots p_{d_h}(s)$ (5.5.30)

where the zeros of $\{p_{d_i}(s) : i=1,2,\ldots,\ell\}$ are

all in the left half-plane.

In these situations it is possible to combine conditions (1a)
and (4), of statement 1, into the single equivalent condition

$$\sum_{i=1}^{\ell} N(\overset{*}{g}_i, -k) - \sum_{i=1}^{\ell} N(\overset{*}{g}_i, 0) = - \sum_{i=1}^{h} P_{G_i}$$ (5.5.31)

The necessity and sufficiency of condition (5.5.31) for closed-loop stability when conditions (2) and (3) are satisfied is proved as follows.

From equations (4.2.37) and (4.2.39) we have that

$$\sum_{i=1}^{h} P_{G_i} = e + \sum_{i=1}^{\ell} P_{f_i} + P_d \qquad (5.5.32)$$

But the poles of $\{f_i(s): i=1,2,\ldots,\ell\}$ are the same as the poles of $\{g_i(s): i=1,2,\ldots,\ell\}$ which in turn are the same as the zeros of $\{g_i^*(s): i=1,2,\ldots,\ell\}$, and therefore equation (5.5.32) can be rewritten as

$$\sum_{i=1}^{h} P_{G_i} = e + \sum_{i=1}^{\ell} Z_{g_i^*} + P_d \qquad (5.5.33)$$

Equation (5.5.33) can now be combined with equation (5.5.19) to give

$$\sum_{i=1}^{\ell} N(r_i^*,0) - \sum_{i=1}^{\ell} N(g_i^*,0) = \sum_{i=1}^{\ell} Z_{r_i^*} + e + P_d - \sum_{i=1}^{h} P_{G_i}$$

$$(5.5.34)$$

which using equation (5.5.24) becomes

$$\sum_{i=1}^{\ell} N(g_i^*-k) - \sum_{i=1}^{\ell} N(g_i^*,0) = \sum_{i=1}^{\ell} Z_{r_i^*} + e + P_d - \sum_{i=1}^{h} P_{G_i}$$

$$(5.5.35)$$

To establish the necessity of condition (5.5.31) suppose that

$$\sum_{i=1}^{\ell} N(g_i^*,-k) - \sum_{i=1}^{\ell} N(g^*,0) \neq -\sum_{i=1}^{h} P_{G_i}$$

then from equation (5.5.35) this implies that

$$\sum_{i=1}^{\ell} Z_{r_i^*} \neq 0, \text{ or } e \neq 0, \text{ or } P_d \neq 0$$

or any combination of these, and thus that the system will be closed-loop unstable. Hence we conclude that condition (5.5.31) is necessary for closed-loop stability.

For sufficiency suppose that

$$\sum_{i=1}^{\ell} N(\overset{*}{g_i}, -k) - \sum_{i=1}^{\ell} N(\overset{*}{g_i}, 0) = - \sum_{i=1}^{h} P_{G_i}$$

then from equation (5.5.35) we must have that

$$\sum_{i=1}^{\ell} Z^*_{r_i} = e = P_d = 0$$

and hence the system is stable providing conditions (2) and (3) are satisfied. Thus the sufficiency of condition (5.5.31) is established.

Therefore to complete the proof of statement 2 of the stability criterion all that is left is to show the equivalence between conditions (1a) and (1b). To do this we will show that when conditions (2) and (3) are satisfied, condition (1b) i.e.

$$\sum_{i=1}^{\ell} N(\overset{*}{r_i}, 0) = \sum_{i=1}^{\ell} N(\overset{*}{g_i}, -k) = - \sum_{i=1}^{h} Z_{G_i} \qquad (5.5.36)$$

where Z_{G_i} is the number of right half-plane poles of $G_i(s)$, is necessary and sufficient for closed-loop stability.

From equation (5.5.15), which is valid for $\{j=1,2,\ldots,\ell\}$,

$$\sum_{i=1}^{\ell} N(\overset{*}{r_i}, 0) = \sum_{i=1}^{\ell} N(\overset{*}{g_i}, -k) = \sum_{i=1}^{\ell} Z^*_{r_i} - \sum_{i=1}^{\ell} P^*_{r_i}$$

$$(5.5.37)$$

and since the poles of $\{\overset{*}{r_i}(s): i=1,2,\ldots,\ell\}$ are the same as the poles of $\{\overset{*}{g_i}(s): i=1,2,\ldots,\ell\}$ equation (5.5.37) can be rewritten as

$$\sum_{i=1}^{\ell} N(\overset{*}{g_i}, -k) = \sum_{i=1}^{\ell} Z^*_{r_i} - \sum_{i=1}^{\ell} P^*_{g_i} \qquad (5.5.38)$$

If the subsystems $\{G_i(s): i=1,2,\ldots,h\}$ are each square then $P_x (=P_d)$ is the number of right half-plane zeros (or poles) which are lost through pole-zero cancellations when $G(s)$ is formed. Therefore if the number of right half-plane zeros of $G(s)$ is given $\left[\text{equation (3.3.36)}\right]$ by

$$Z_G = e + \sum_{i=1}^{\ell} Z_{g_i} \qquad (5.5.39)$$

then we must have

$$\sum_{i=1}^{h} Z_{G_i} = e + \sum_{i=1}^{\ell} Z_{g_i} + P_d \qquad (5.5.40)$$

or, since the poles of $\overset{*}{g}_i(s)$ are identically equal to the

zeros, of $g_i(s)$,

$$\sum_{i=1}^{h} Z_{G_i} = e + \sum_{i=1}^{\ell} P_{g_i}^* + P_d \qquad (5.5.41)$$

Consequently equations (5.5.38) and (5.5.41) can be combined

to give

$$\sum_{i=1}^{\ell} N(\overset{*}{g}_i,-k) = \sum_{i=1}^{\ell} Z_{r_i}^* + e + P_d - \sum_{i=1}^{h} Z_{G_i}$$

$$(5.5.42)$$

To establish the necessity of condition (5.5.36) suppose

that

$$\sum_{i=1}^{\ell} N(\overset{*}{g}_i,-k) \neq - \sum_{i=1}^{h} Z_{G_i}$$

then from equation (5.5.42) this implies that

$$\sum_{i=1}^{\ell} Z_{r_i}^* \neq 0, \text{ or } e \neq 0, \text{ or } P_d \neq 0$$

or any combination of these and thus that the system will be

closed-loop unstable. Hence we conclude that condition

(5.5.42) is necessary for closed-loop stability.

For sufficiency suppose that

$$\sum_{i=1}^{\ell} N(\overset{*}{g}_i,-k) = - \sum_{i=1}^{h} Z_{G_i}$$

then from equation (5.5.42) we must have that

$$\sum_{i=1}^{\ell} Z_{r_i}^* = e = P_d = 0$$

and hence the system is closed-loop stable, providing

conditions (2) and (3) are satisfied. Thus the sufficiency

of condition (5.5.36) is established.

This completes the proof of statement 2 of the generalized inverse Nyquist stability criterion.   ■

5.6   Example

To illustrate the stability criteria consider the system example used in chapters 3 and 4 where

$$G(s) \ = \ G_1(s) = \frac{1}{1.25\,(s+1)\,(s+2)} \begin{bmatrix} s-1 & s \\ -6 & s-2 \end{bmatrix}$$

The pole and zero polynomials for the open-loop gain matrix G(s) are

$$p_G(s) \ = \ (s+1)\,(s+2) \quad \text{and} \quad z_G(s) = 1$$

so that

$$P_G = 0 \quad \text{and} \quad Z_G = 0.$$

The inverse characteristic gain loci are shown in figure 19.

There are no unobservable and/or uncontrollable modes and therefore either statement 1 or statement 2 of the criterion is directly applicable.   In fact, because we have effectively only one subsystem, there is no difference between the two statements.

Testing the conditions for stability as outlined by the generalized inverse Nyquist stability criterion we find that the closed-loop system is stable for

$$-1.875 < k \leqslant 0$$

$$0 \leqslant k < 1.25$$

and $2.5 < k < \infty$

These bounds on the gain control variable k agree with those obtained in section 4.3 using the generalized Nyquist stability criterion.

It is interesting to note that an inverse Nyquist-like

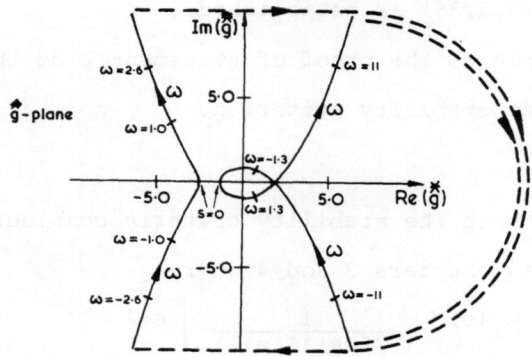

Figure 19. Inverse characteristic gain loci

stability criterion has recently been used by Mees and Rapp
[6] to establish stability criteria for multiple-loop <u>nonlinear</u>
feedback systems.

References

[1] A.L. Whiteley, "Fundamental Principles of Automatic Regulators
and Servo Mechanisms", J. IEE, 5-22, 1947.

[2] H.H. Rosenbrock, "State Space and Multivariable Theory",
Nelson, London, 1970.

[3] A.G.J. MacFarlane, "Dynamical System Models", Harrap, London,
1970.

[4] A.G.J. MacFarlane, "Return-difference and return-ratio matrices
and their use in analysis and design of multivariable
feedback control systems", Proc. IEE, 117, 2037-2049, 1970.

[5] E. Hille, "Analytic Function Theory", Vol. 1, Ginn and Go.,
U.S.A., 1959.

[6] A.I. Mees and P.E. Rapp, "Stability criteria for multiple-
loop nonlinear feedback systems", Proc. IFAC Fourth Multivariable
Technological Systems Symposium, Fredericton, Canada, 1977.

6.   Multivariable root loci

The Evans' root locus approach [1;2;3] is a well established graphical technique used in the analysis and design of linear time-invariant single-input single-output feedback systems for estimating the system closed-loop poles as a function of the gain control variable.   Its generalization to the multivariable case has caused considerable interest and steps towards this end have been made with regard to the asymptotic behaviour of the characteristic frequency loci [4; 5] , and the angles of departure and approach of the characteristic frequency loci [6] ; both from a state-space point of view.   In this chapter a method is developed, using well established results in algebraic function theory [7;8;9], which allows the asymptotic behaviour and the angles of departure and approach of the characteristic frequency loci to be determined from a Laplace transfer function matrix description of the system.

It is also shown how the method can be utilised to find the asymptotic behaviour of the optimal closed-loop poles of a multivariable time-invariant linear regulator as the weight on the input in the performance criterion approaches zero [10].

The method seems particularly useful for systems with small numbers of inputs since the calculations are then simple enough to be carried out by hand.

6.1   Theoretical background

For the general feedback configuration of figure 3 it has been shown in sub-section 3.3-4 that the characteristic frequency loci (multivariable root loci) are the $180°$ phase

contours of the characteristic gain function $g(s)$, where $g(s)$

is defined via the equation

$$\Delta(g,s) \triangleq \det\left[gI_m - G(s)\right] = 0 \qquad (6.1.1)$$

or the equation

$$\Phi(g,s) = b_o(s)\Delta(g,s) = 0 \qquad (6.1.2)$$

where we have multiplied through by the least common denominator

of the rational coefficients in s.

For simplicity of exposition, and because this is in any case

the usual situation for transfer function matrices arising

from practical situations, it is assumed that $\Delta(g,s)$ is

irreducible over the field of rational functions in s.   The

characteristic gain function $g(s)$ is related to the gain

control variable k via the expression

$$g = -\frac{1}{k} \qquad (6.1.3)$$

so that substituting for g in equation (6.1.2) we have

$$\Phi(-k^{-1},s) = k^{-m}\,\Gamma(k,s) = 0 \qquad (6.1.4)$$

and solutions of

$$\Gamma(k,s) = 0 \qquad (6.1.5)$$

for s in terms of positive real k, determine the dependence

of the closed-loop poles on the gain control variable k.

Therefore, apart from possible single-point loci, the graphical

description of this dependence constitutes the characteristic

frequency loci of the given system.   Note that if there are

no single-point loci then equation (6.1.5) is directly

equivalent to

$$CLCP(s) \triangleq \det\left[sI_n - A_c\right] = \det\left[sI_n - S(-k^{-1})\right] = 0 \qquad (6.1.6)$$

and the characteristic frequency loci have n branches.

Each of the n branches of the characteristic frequency loci can in theory be represented in the form

$$s_i(k) = u_i(k) + jv_i(k) \quad i=1,2,\ldots,n \quad (6.1.7)$$

where $j=\sqrt{-1}$, and the subscripts i are labels for the various branches. The tangent to the rth branch of the loci at a point $s_o=s_r(k_o)$ is defined as the limiting position of the straight line through s_o and another point $s_1=s_r(k_o+\delta k)$ as s_1 approaches s_o along the branch, that is as $\delta k \to 0$. Now the complex number s_1-s_o can be represented by the vector from s_o to s_1 (figure 20), and the vector corresponding to $(s_1-s_o)\div\delta k$, where $\delta k>0$, has the same direction as that vector.

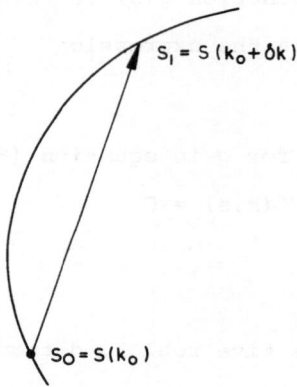

Figure 20. To derive a formula for the tangent to the characteristic frequency loci

It follows therefore that the vector corresponding to

$$\dot{s}_r(k_o) = \left.\frac{ds_r(k)}{dk}\right|_{k_o} = \lim_{\delta k \to 0} \frac{s_1 - s_o}{\delta k}$$

$$= \lim_{\delta k \to 0} \frac{s_r(k_o + \delta k) - s_r(k_o)}{\delta k} \quad (6.1.8)$$

is tangent to the branch at s_o and the angle θ between this
vector and the positive real axis is given by

$$\theta = \text{argument } \{\dot{s}_r(k_o)\} \quad (6.1.9)$$

Formula (6.1.9) is fundamental to the determination of
the asymptotic behaviour of the characteristic frequency
loci, and the angles of departure and approach of the loci.
For k=0, equation (6.1.3) implies that the characteristic
frequency loci start at poles of the open-loop system, and
therefore the angle of departure of a branch of the loci from
a pole is given by formula (6.1.9) with $k_o = 0$. For k=∞,
equation (6.1.3) implies that the characteristic frequency
loci terminate at system zeros.   Therefore for a branch
of the loci terminating at a finite zero, formula (6.1.9),
with $k_o = \infty$, gives the angle of approach of the branch to the
zero.   If a branch terminates at an infinite zero then formula
(6.1.9), with $k_o = \infty$, gives the angle that the asymptote to
the branch makes with the positive real axis.

Formula (6.1.9) does not at first seem to be useful,
because expressions for the separate branches of the characteristic
frequency loci cannot in general be found explicitly.   However,
in algebraic function theory [7;8;9], there exists a method
for the practical construction of ˙ series representations
for the branches of an algebraic function in the neighbourhood
of a given point.   The method consists of repeatedly using
a "Newton diagram" to find the next most significant terms
in the series.   Therefore, by using the Newton diagram just
once, approximations can be obtained for the branches of the

characteristic frequency loci in the vicinity of a pole
or zero (finite or infinite), of the form

$$s_r(k) \simeq a + bk^\alpha \qquad (6.1.10)$$

where a and b are complex numbers, α is a rational real
number, and when α is fractional the principal root of k is
understood.   For the negative feedback configuration under
consideration k is real and positive, $\alpha k^{\alpha-1}$ will always be
real, and therefore applying formula (6.1.9) to the approximation
(6.1.10) we have

$$\theta = \text{argument } \{b\} \quad \pm \tau 180° \qquad (6.1.11)$$

where

$$\tau \begin{cases} = 0 \text{ when } \alpha > 0 \\ \\ = 1 \text{ when } \alpha < 0 \end{cases}$$

as a result of the differentiation with respect to k.

In the following sections it is shown how to obtain
approximations of the form shown in equation (6.1.10) in
the vicinity of a pole or zero, and hence to determine the
asymptotic behaviour of the characteristic frequency loci
and also the angles of departure and approach of the loci.

6.2   Asymptotic behaviour

The Newton diagram is a graphical construction which
can be used to find each of the most significant terms in the
series representations for the particular branches of an
algebraic function q(v), in the vicinity of the origin,
which approach zero as v approaches zero.   Therefore, whether
finding the asymptotic behaviour or the angles of departure
and approach of the loci, our first aim is always to reduce
the problem, by a change of variables in the characteristic

equation, to one of finding approximations for branches of
an algebraic function which approach zero as the independent
variable approaches zero.

To determine the asymptotic behaviour of the characteristic
frequency loci we need to obtain an approximation to the branch
(or branches) of the loci about the point $s=\infty$, as k approaches
infinity.   For this purpose we will put

$$s = z^{-1} \qquad\qquad\qquad\qquad (6.2.1)$$

in the characteristic equation (6.1.2) to obtain

$$\Phi(g,s) = \Phi(g,z^{-1}) = z^{-q} \ \Psi(g,z) \qquad\qquad (6.2.2)$$

where $q \leqslant n$ is the number of poles of g(s), so that in any
neighbourhood of the value z = 0 (the point z = 0 itself is
excluded from the region) the equation $\Phi(g,s)=0$ is equivalent
to the equation

$$\Psi(g,z) = \Sigma\psi_{xy}g^{x}z^{y} = 0 \qquad\qquad\qquad (6.2.3)$$

For a strictly proper system, that is one where D is zero,
or if D is singular

$$\psi_{oo} = 0 \qquad\qquad\qquad\qquad (6.2.4)$$

otherwise ψ_{oo} is non-zero which, as we shall see later from
the Newton diagram, corresponds to there being no asymptotic
behaviour.   Note that if D is non-singular g(s) has the same
number of finite zeros as poles and therefore we would not
expect any closed-loop poles to approach infinity.

The next step is to construct the Newton diagram for
$\Psi(g,z)$, from which approximations of the form

$$z \simeq cg^{\mu} \qquad\qquad\qquad\qquad (6.2.5)$$

can be obtained, where c is a complex number, and μ a rational
real number.   The procedure for constructing the appropriate
Newton diagram is as follows [8].

In a (g,z)-plane we plot the points (x,y) for which

$\psi_{xy} \neq 0$ in equation (6.2.3). As an example, the points for

$$(1-z+2z^2-25z^3+29z^4)g^2 + (z-22z^2+199z^3-210z^4)g$$

$$-(33z^3-594z^4) = 0 \qquad (6.2.6)$$

are shown in figure 21. A straight line is then made to coincide with the horizontal axis and rotated clockwise about the smallest g-axis point P_0 (the point $(2,0)$ on figure 21) until this line passes through another point P_1 of our net. A straight line is then drawn between the points P_0 and P_1. A horizontal line through P_1 , and pointing away from P_1 towards the vertical z-axis, is then rotated clockwise about the point P_1 until it passes through another point P_2 of our net. A straight line is then drawn between the points P_1 and P_2. The procedure is repeated until the vertical z-axis is reached. The complete Newton diagram for equation (6.2.6) is shown in figure 22.

The tangent of the acute angle which the straight line P_0P_1 makes with the vertical z-axis determines the first possible value of μ , and the tangent of the acute angle which the straight line P_1P_2 makes with the verticle z-axis determines the second possible value of μ , etc. For the Newton diagram of figure 22 we have $\mu_1 = 1$ and $\mu_2 = \frac{1}{2}$.

For a particular exponent μ_t there may be several approximations of the form

$$z_{it} \overset{\sim}{=} c_{it} g^{\mu_t} \qquad (6.2.7)$$

Note that if some root of g is implied by μ_t i.e. μ_t is a fraction, then it is understood that the principal root is being considered. To determine the coefficients c_{it} it

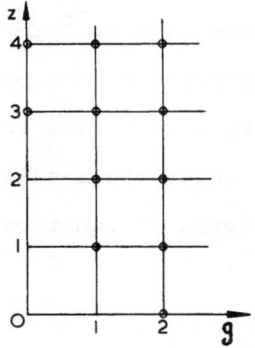

Figure 21. Points of the Newton diagram
for equation (6·2·6)

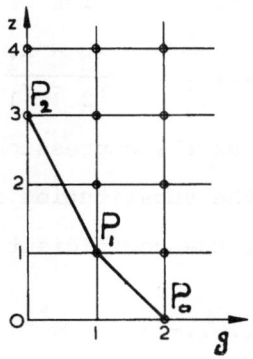

Figure 22. Complete Newton diagram
for equation (6·2·6)

is necessary to substitute $z = cg^{\mu_t}$ in the terms of equation (6.2.3) corresponding to the points of our net lying on the link $P_{t-1} P_t$, to equate to zero the result of the substitution and to solve the resulting equation. Let the sum of the relevant terms of equation (6.2.3) have the form

$$\psi_{x_\sigma y_\sigma} g^{x_\sigma} z^{y_\sigma} + \ldots\ldots + \psi_{x_1 y_1} g^{x_1} z^{y_1} \qquad (6.2.8)$$

where $y_\sigma > \ldots\ldots\ldots > y_1$, and σ is a positive integer less than the number of infinite zeros of the system. Then because the terms correspond to the same link we have

$$\frac{x_\sigma - x_1}{y_\sigma - y_1} = \frac{x_{\sigma-1} - x_1}{y_{\sigma-1} - y_1} = \ldots\ldots = \frac{x_2 - x_1}{y_2 - y_1} = \mu_t$$

Therefore all the terms of the expression (6.2.8) become similar as a result of the substitution $z = cg^{\mu_t}$, and hence for the determination of the coefficient c_{it} we obtain an equation of the form

$$\psi_{x_\sigma y_\sigma} c^{y_\sigma} + \ldots\ldots + \psi_{x_1 y_1} c^{y_1} = 0 \qquad (6.2.9)$$

or dividing by c^{y_1} $(c \neq 0)$ we obtain

$$\psi_{x_\sigma y_\sigma} c^{y_\sigma - y_1} + \ldots + \psi_{x_1 y_1} = 0 \qquad (6.2.10)$$

which has $y_\sigma - y_1$ solutions. Note that $y_\sigma - y_1$ is the difference between the ordinates of the points P_t and P_{t-1}. Consequently it is equal to the projection of the link $P_{t-1} P_t$ on to the ordinate axis. This is a useful fact since it enables us to see at a glance how many coefficients there are corresponding to the exponent μ_t. Also the ordinate associated with the point where the final link reaches the vertical axis must give the total number of coefficients for all the approximations considering all exponents. This vertical axis point on the final link is therefore the number of infinite zeros of the system.

Having obtained the approximations in the form of equation (6.2.7) it is a simple matter of substitution from equations (6.1.3) and (6.2.1) to obtain the required form

$$s_{it} \overset{\sim}{=} b_{it} \, k^{\alpha_t} \qquad\qquad (6.2.11)$$

representing approximations to the characteristic frequency loci about s=∞, as k→∞. If we now apply formula (6.1.9) we obtain the angles which the asymptotes make with the positive real axis.

The asymptotes, as will be shown in sub-section 6.2-1, group into Butterworth configurations. Each pattern has a common intercept which has been termed the "multivariable pivot" by Kouvaritakis and Shaked [4] who also gave a method for calculating the pivot given the state space description of the system. In appendix 6 it is shown how the multivariable pivots can be derived from the characteristic equation $\Delta(g,s) = 0$.

6.2-1   Butterworth Patterns

In this sub-section it is proved that the closed-loop poles that go off to infinity do so along asymptotes which group into several Butterworth configurations. The proof is based on well established results in algebraic function theory [8] .

Let us consider the closed-loop characteristic polynomial defined by equation (6.1.5) i.e.

$$\Gamma(k,s) = 0 \qquad\qquad (6.2.12)$$

in which we allow k to be complex. Also suppose that at a point k_c the algebraic function s(k) defined by equation (6.2.12) has a p- fold root s_c, and let the n branches of s(k) in a neighbourhood N of k_c be $\{s_i(k):i=1,2,\ldots,p,\ldots,n\}$ where $\{s_i(k_c)=s_c:i=1,2,\ldots,p\}$.

If we analytically continue any one of the branches $\{s_i(k) : i=p+1,\ldots,n\}$ around k_c in N, then after one encirclement we return to the original function. If, however, we analytically continue $s_1(k)$ around the critical point k_c then after one encirclement we obtain one of the functions $\{s_i(k) : i=2,3,\ldots,p\}$, $s_2(k)$ say, and after another encirclement one of the functions $\{s_i(k) : i=3,4,\ldots p\}$, $s_3(k)$ say. After a finite number ν_1, of encirclements we return to the original function $s_1(k)$. In this case we shall say that the functions

$$s_1(k),\ldots, s_{\nu_1}(k) \qquad\qquad (6.2.13)$$

constitute a cyclical system of branches of the algebraic function s(k) (in the neighbourhood N of k_c).

If $\nu_1 < p$ we take the function $s_{\nu_1+1}(k)$ and analytically continue it around k_c in N. Repeating the previous arguments we obtain a second cyclical system of branches, namely

$$s_{\nu_1+1}(k),\ldots, s_{\nu_2}(k) \qquad\qquad (6.2.14)$$

Finally all the functions $s_1(k),\ldots,s_p(k)$ will be sorted into cyclical systems.

The functions of each cyclical system pass consecutively into one another according to the cyclical law when k goes

round the multiple point k_c.   Hence every cyclical system
of branches defines in the neighbourhood N of k_c an   analytic
function.   The number of values given by this analytic
function at a point in the neighbourhood N is equal to the
number of branches comprising the cyclical system being
considered.   When a single branch comprises a cyclical system,
the corresponding analytic function will be single-valued
and identical with this branch.

Suppose the neighbourhood N is replaced by a pile of
n such regions, one for each of the functions $\{s_i(k) : i=1,2,\ldots,n\}$
and let a point on the ith sheet represent the pair $\left[k, s_i(k)\right]$.
Also suppose that each neighbourhood N is circular and has a
radial line cut from the critical point k_c to its periphery,
as shown in figure 23.   Then the neighbourhoods can be joined
together along the edges of the cuts so that if we analytically
continue the jth branch (j<p) around k_c on the jth sheet,
we move after one encirclement across the cut into the (j+1)th
branch on the (j+1)th sheet, etc. eventually returning to
the jth branch.   The pile of neighbourhoods obviously
falls into cycles containing one or more sheets as shown in
figure 24, and each cycle is a domain for one of the analytic
functions defined in the neighbourhood N of k_c.

To denote an analytic function determined by a cyclical
system we will use a Roman   numerical subscript.   For example
the analytic function corresponding to the cyclical system
$s_1(k),\ldots, s_{\nu_1}(k)$ will be denoted by the symbol $s_I(k)$.

When we consider the asymptotic behaviour of the

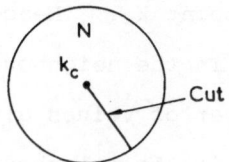

Figure 23. Neighbourhood N of the
critical point k_c

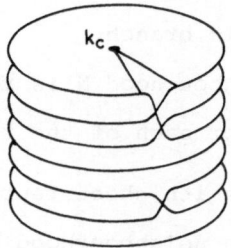

Figure 24. Cycles of neighbourhoods for
the critical point k_c

characteristic frequency loci, we are concerned with the
branches of the algebraic function s(k) which approach
infinity, as k approaches infinity. Suppose we have p
infinite branches for k=∞, then by the previous arguments
the p branches will be arranged into several cyclical
systems. Each cyclical system is defined by an analytic
function, the form of which is given by the following
theorem [8, page 39] .

Theorem 4

   The infinite branches of the algebraic function s(k),
constituting a cyclical system in the neighbourhood of its
critical point $k_c=∞$, in their totality constitute an analytic

function given by a series of the form

$$s_I = k^{\frac{\lambda}{\nu}} (b_o + b_1 k^{\frac{-1}{\nu}} + b_2 k^{\frac{-2}{\nu}} + \ldots) \qquad (6.2.15)$$

where ν is the number of branches in the cycle, and λ is a positive integer.

Consequently the infinite branches corresponding to a particular cyclical system in the neighbourhood of $k=\infty$, can be approximated by an analytic function of the form

$$s_I(k) \stackrel{\sim}{-} b_o k^{\frac{\lambda}{\nu}} \qquad (6.2.16)$$

The characteristic frequency loci are values of the algebraic function $s(k)$ for k going from zero to infinity along the positive real axis. Therefore the infinite branches of the frequency loci will occur in cyclical systems which can be approximated by equations of the form (6.2.16). Taking the ν roots of k in equation (6.2.16) we find that the asymptotes corresponding to a cyclical system arrange themselves into a Butterworth configuration of order ν ; that is, we have ν asymptotes equally spaced by angles of $(\frac{360}{\nu})^o$, in the complex plane.

In general, as already discussed, there will be several cyclical systems of branches in the neighbourhood of $k = \infty$, and therefore the asymptotes of the characteristic frequency loci will arrange themselves into several Butterworth configurations.

6.3   Angles of departure and approach

To determine the angles of departure and approach of the characteristic frequency loci we need to obtain approximations to the branches of the loci at the poles and finite zeros

of the system.   We will first consider the angles of
departure from the open-loop poles of the system.

Suppose for the characteristic equation

$$\Phi(g,s) = 0 \qquad\qquad (6.3.1)$$

we have a pole (or multiple pole) at $s = \beta$.   We will make
$s'=s-\beta$ the new independent variable for equation (6.3.1)
and also make the substitution $g = d^{-1}$, so that equation
(6.3.1) becomes

$$\Phi(g,s) = \Phi(d^{-1},s'+\beta) = d^{-m}\Xi(d,s') = 0 \qquad (6.3.2)$$

The situation we are examining now reduces to the case where
$d = s' = 0$, and in the neighbourhood of $s = \beta$ (excluding β
itself) equation $\Phi(g,s) = 0$ is equivalent to the equation

$$\Xi(d,s') = \Sigma\xi_{xy}d^x s'^y = 0 \ , \quad \xi_{oo} = 0 \qquad (6.3.3)$$

If we construct the Newton diagram for the equation
$\Xi(d,s') = 0$ an approximation (or approximations in the case of
multiple poles) of the form

$$s' \cong ed^{\omega} \qquad\qquad (6.3.4)$$

is obtained, where e is a complex number and ω a rational
real number.   From equation (6.1.3), the change of variable,
and the substitution $g = d^{-1}$, we therefore arrive at an
approximation to the branch (or branches) of the characteristic
frequency loci departing from the pole β , in the form

$$s \cong \beta + b_d k^{\mu d} \qquad\qquad (6.3.5)$$

If we now apply formula (6.1.9) the angle of departure θ_d is
given as

$$\theta_d = \text{argument } \{b_d\} \qquad\qquad (6.3.6)$$

We will now consider the angles of approach to the finite

zeros of the system.   Suppose for the characteristic
equation (6.3.1) we have a zero (or multiple zero) at $s = \gamma$.
We will take $s' = s-\gamma$ as the new independent variable so
that the equation becomes

$$\Phi(g,s) = \Phi(g,s'+\gamma) = \chi(g,s') = 0 \qquad (6.3.7)$$

The situation reduces to the case where $g = s' = 0$, and in
the neighbourhood of $s = \gamma$ (excluding γ itself) equation
$\Phi(g,s) = 0$ is equivalent to the equation

$$\chi(g,s') = \Sigma \times_{xy} g^x s'^{\underline{y}} = 0 \ , \times_{oo} = 0 \qquad (6.3.8)$$

If we construct the Newton diagram for $\chi(g,s')$ an
approximation of the form

$$s' \overset{\sim}{-} pg^{\eta} \qquad (6.3.9)$$

is obtained where p is a complex number and η is a rational
real number.   From equation (6.1.3) and the change of variable
we therefore arrive at an approximation to the branch (or
branches) of the characteristic frequency loci approaching
the zero $s = \gamma$, in the form

$$s \overset{\sim}{-} \gamma + b_a k^{\mu} a \qquad (6.3.10)$$

If we now apply formula (6.1.9) the angle of approach θ_a is
given as

$$\theta_a = \text{argument } \{b_a\} \pm 180° \qquad (6.3.11)$$

Note that μ_a will always be negative and hence the presence
of the 180° term in formula (6.3.11).   Also note that this
definition for the angle of approach differs by 180° from the
usual definition which is the direction you would look, when
positioned at the zero, to see the locus arrive.

6.4   Example 1

Consider an open-loop gain matrix

$$G(s) = \frac{1}{s^4+5s^3-2s^2-44s+40} \begin{bmatrix} 3s^3+4s^2-156s+464 & 8s^2-24s+16 \\ s^3+79s^2+44s-868 & -4s^3-4s^2+40s-32 \end{bmatrix}$$

which has a characteristic equation

$$\Phi(g,s) = (s^4+5s^3-2s^2-44s+40)g^2 + (s^3+116s-432)g$$
$$-12(s^2-2s+2) = 0$$

(a)   To determine the asymptotic behaviour

Putting $s = z^{-1}$ in the characteristic equation we obtain

$$\Psi(g,z) = (1+5z-2z^2-44z^3+40z^4)g^2 + (z+116z^3-432z^4)g$$
$$-12(z^2-2z^3+2z^4) = 0$$

The Newton diagram for $\Psi(g,z)$ is shown in figure 25 from which we obtain $\mu = 1$, and hence the approximation

$$z \overset{\sim}{=} cg$$

The coefficient c is calculated (using the method described in section 6.2) to have the values $-\frac{1}{4}$ and $\frac{1}{3}$.   Therefore, resubstituting for z and $g=-k^{-1}$, we have the following approximations for the characteristic frequency loci

$$s \overset{\sim}{=} 4k \quad \text{and } s \overset{\sim}{=} -3k \quad \text{as } k \to \infty$$

Two branches of the characteristic frequency loci therefore move off to infinity at angles of 0° and 180° to the positive real axis.

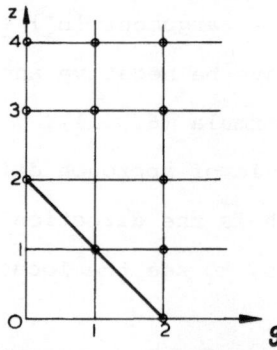

Figure 25. Newton diagram for $\Psi(g,z)$

(b)   To find the angles of departure

The system has four open-loop poles

$s = 1$, $s = 2$, $s = -4+2j$, and $s = -4-2j$

Pole at $s = 1$

Putting $s' = s-1$ and $g = d^{-1}$ in the characteristic

equation we obtain

$$\Xi_1(d,s') = (s'^4+9s'^3+19s'^2-29s') + (s'^3+3s'^2+119s'-315)d$$
$$- 12(s'^2+1)d^2 = 0$$

The Newton diagram for $\Xi_1(d,s')$ is shown in figure 26 from

which we obtain $\omega_1 = 1$, and hence the approximation

$$s' \overset{\sim}{-} e_1 d$$

The coefficient e_1 is calculated (using the method described

in section 6.2) to have the value -10.86, resulting in the

following approximation to the characteristic frequency loci

$$s \overset{\sim}{-} 1 + 10.86 \ k$$

about the pole $s = 1$.   Therefore the angle of departure from

the pole $s = 1$ is $0°$.

Pole at $s = 2$

Putting $s' = s-2$ and $g = d^{-1}$ in the characteristic

equation we obtain

$$\Xi_2(d,s') = (s'^4+13s'^3+52s'^2+40s') + (s'^3+6s'^2+128s'-192)d$$
$$- 12(s'^2+2s'+2)d^2 = 0$$

The Newton diagram for $\Xi_2(d,s')$ is shown in figure 27 from

which we obtain $\omega_2 = 1$, and hence the approximation

$$s' \overset{\sim}{-} e_2 d$$

The coefficient e_2 is calculated to have the value 4.8,

resulting in the following approximation to the characteristic

frequency loci

$$s \overset{\sim}{=} 2 - 4.8\,k$$

about the pole $s = 2$.   Therefore the angle of departure from
the pole $s = 2$ is $180°$.

Pole at $s = -4+2j$

Putting $s' = s+4-2j$ and $g = d^{-1}$ in the characteristic
equation we obtain

$$\Xi_3(d,s') = \left[s'^4+(-11+j\,8)\,s'^3+(10-j\,60)\,s'^2+(88+j\,104)\,s'\right]$$

$$+\left[s'^3+(-12+j\,6)\,s'^2+(152-j\,48)\,s'+(-912+j\,320)\right]d$$

$$-12\left[s'^2+(-10+j\,4)\,s'+(22-j\,20)\right]d^2 = 0$$

The Newton diagram for $\Xi_3(d,s')$ is shown in figure 28 from
which we obtain $\omega_3 = 1$, and hence the approximation

$$s' \overset{\sim}{=} e_3 d$$

The coefficient e_3 is calculated to have the value

$$\frac{912 - j320}{88 + j104}$$

resulting in the following approximation to the characteristic
frequency loci

$$s \overset{\sim}{=} -4+2j + \frac{(-912+j320)}{(88+j104)}\,k$$

about the pole $s = -4+2j$.   Therefore the angle of departure
from the pole $s = -4+2j$ is

$$\text{argument } \left\{\frac{-912+j320}{88+j104}\right\} = 110.9°$$

Pole at $s = -4-2j$

By symmetry the angle of departure from the pole
$s = -4-2j$ is $-110.9°$.

(c)   To find the angles of approach

The system has two finite zeros

$$s = 1+j \quad \text{and} \quad s = 1-j$$

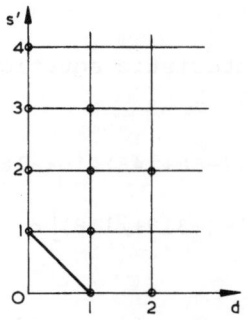

Figure 26. Newton
diagram for $\Xi_1(d,s')$

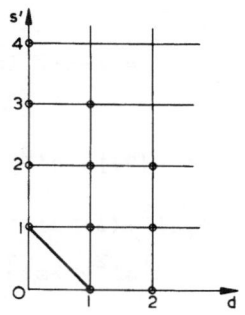

Figure 27. Newton
diagram for $\Xi_2(d,s')$

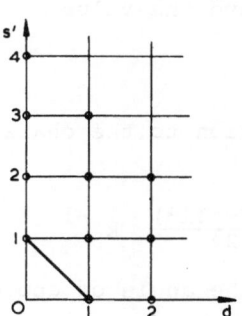

Figure 28. Newton
diagram for $\Xi_3(d,s')$

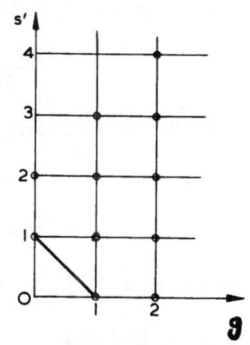

Figure 29. Newton
diagram for $X(g,s')$

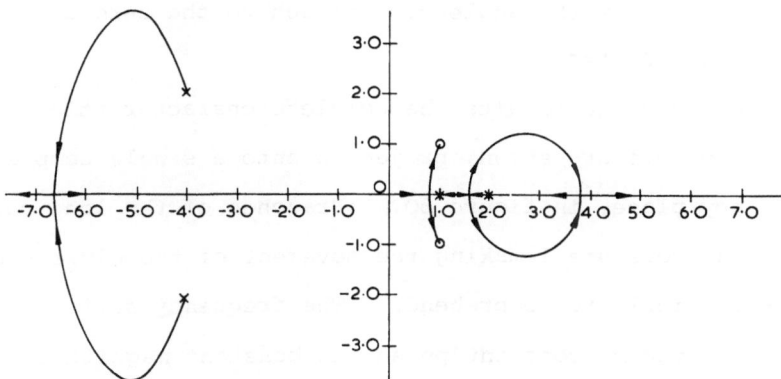

Figure 30. Complete characteristic frequency loci

Zero at s = 1+j

Putting s' = s-1-j in the characteristic equation we obtain

$$\chi(g,s') = \left[s'^4 + (9+j4)s'^3 + (13+j29)s'^2 + (-58+j46)s' + (-18-j38) \right] g^2$$
$$+ \left[s'^3 + (3+j3)s'^2 + (116+j8)s' + (-318+j118) \right] g$$
$$- 12\left[s'^2 + j2s' \right] = 0$$

The Newton diagram for $\chi(g,s')$ is shown in figure 29 from which we obtain $\eta = 1$, and hence the approximation

$$s' \overset{\sim}{=} pg$$

The coefficient p is calculated to have the value

$$\frac{-318 + j118}{j24}$$

resulting in the following approximation to the characteristic frequency loci

$$s \overset{\sim}{=} 1 + j + \frac{(318-j118)}{j24} \, k^{-1}$$

about the zero s = 1+j.   Therefore the angle of approach to the zero s = 1+j is

$$\text{argument } \left\{ \frac{318-j118}{j24} \right\} \pm 180° \equiv 69.64°$$

Zero at s = 1-j

By symmetry the angle of approach to the zero at s = 1-j is -69.64°.

To check the results the complete characteristic frequency loci are shown, projected onto a single complex frequency plane, in figure 30.   Branches of the loci coincide about the pole s = 1 making the movement of the closed-loop poles difficult to comprehend.   The frequency surface characterized by constant phase and constant magnitude contours of g(s) is therefore shown in figures 31(a), (b)

121

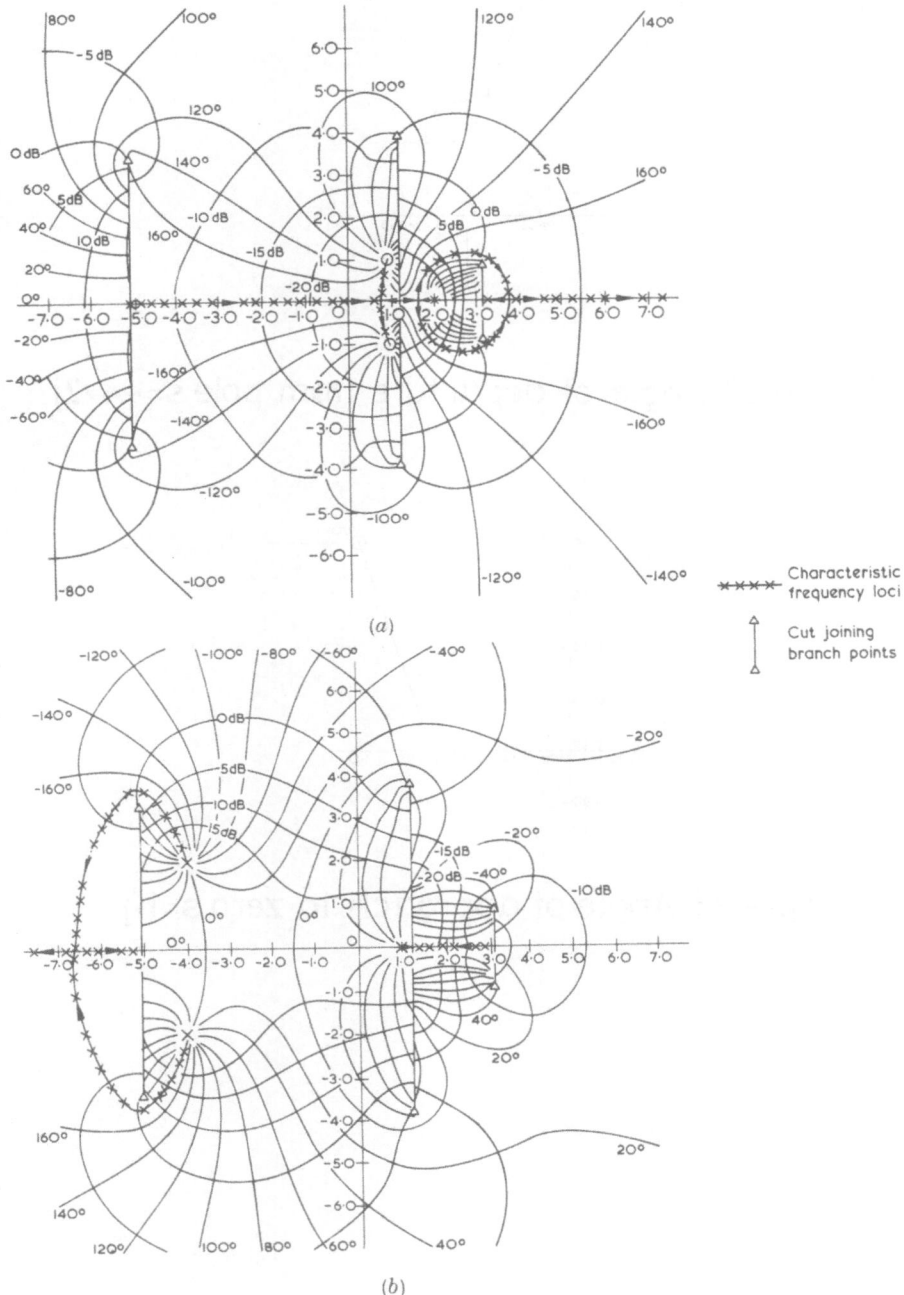

Characteristic
frequency loci

Cut joining
branch points

(a)

(b)

Figure 31. Frequency surface (a) sheet 1 (b) sheet 2

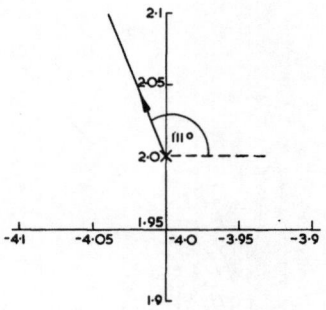

Figure 32. Angle of departure from pole s=−4+2j

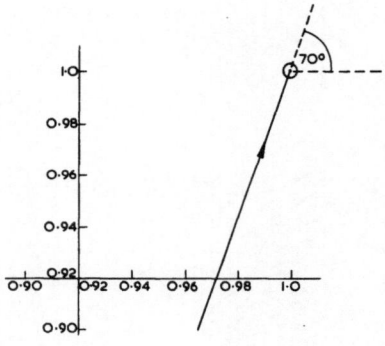

Figure 33. Angle of approach to zero s=1+j

to give a clearer description of the characteristic frequency
loci.   Small regions in the neighbourhood of the pole
s = -4+2j and the zero s = 1+j are shown in figures 32 and
33 to verify the calculated angles of departure and approach.

6.5 Asymptotic behaviour of optimal closed-loop poles

In this section the "Newton diagram" approach is used
to determine the asymptotic behaviour of the optimal closed-
loop poles of a multivariable time-invariant linear regulator,
as the weight on the input in the performance criterion
approaches zero.   The method is based on an association of
the optimal characteristic frequency loci with the branches
of an appropriate algebraic function.   Although the procedure
is equivalent to that given by Kwakernaak [11] the essential
simplicity of the approach is emphasized in the setting of
algebraic function theory.

Consider the stabilizable and detectable time-invariant
linear system

$$\frac{dx(t)}{dt} = A\,x(t) \;+\; Bu(t) \qquad\qquad (6.5.1)$$

$$y(t) = C\,x(t) \qquad\qquad (6.5.2)$$

and the performance criterion

$$V(\infty) = \int_{0}^{\infty}\!\left[y^{T}(t)Qy(t)+\rho u^{T}(t)\,Ru(t)\right]\,dt \qquad (6.5.3)$$

where Q and R are positive definite symmetric matrices,
and the superscript T denotes the transpose of a matrix or
vector.   Then it is well known (e.g. [12]) that the optimal
control action is given by

$$u(t) = -\frac{1}{\rho}\,R^{-1}B^{T}\,Px(t) \qquad\qquad (6.5.4)$$

where P is the unique positive semi-definite solution of

the steady state matrix Riccati equation

$$-PA -A^T P + \frac{1}{\rho}PBR^{-1}B^T P = C^T QC \tag{6.5.5}$$

Kwakernaak [11, equation (16)] has related the closed-loop characteristic polynomial (denoted by $\phi_c(s)$) for the optimal regulator to the open-loop characteristic polynomial (denoted by $\phi_o(s)$) as follows

$$\phi_c(s)\phi_c(-s) = \phi_o(s)\phi_o(-s) \det\left[I_m + \frac{1}{\rho}R^{-1}G^T(-s)QG(s)\right] \tag{6.5.6}$$

where I_m is a unit matrix of order m, the number of system inputs, and

$$G(s) = C(sI - A)^{-1}B \tag{6.5.7}$$

is the open-loop transfer function or gain matrix of the system. It is obvious from (6.5.6) that the poles of the optimal closed-loop system, dependent on the input weighing, are values of s in the left half-plane which satisfy

$$\det\left[I_m + \frac{1}{\rho}H(s)\right] = 0 \tag{6.5.8}$$

where

$$H(s) \triangleq R^{-1}G^T(-s)QG(s) \tag{6.5.9}$$

Consider now the characteristic equation defining the eigenvalues of H(s), that is

$$\Delta(\eta,s) \triangleq \det\left[\eta I_m -H(s)\right] = 0 \tag{6.5.10}$$

or by expanding the determinant

$$\Delta(\eta,s) = \eta^m + a_1(s^2)\eta^{m-1} + \ldots\ldots + a_m(s^2) = 0 \tag{6.5.11}$$

where the coefficients $\{a_i(s^2) \; ; \; i=1,2,\ldots,m\}$ are rational functions in s^2. The coefficients are functions in s^2 because

$$\Delta(\eta,s) = \det\left[\eta I_m - R^{-1}G^T(-s) \, QG(s)\right]$$ (6.5.12)

$$= \det\left[\eta I_m - R^{-\frac{1}{2}}G^T(-s)QG(s)R^{-\frac{1}{2}}\right]$$

and $R^{-\frac{1}{2}}G^T(-s)QG(s)R^{-\frac{1}{2}}$ is para-Hermitian implying [11, appendix A] that $\Delta(\eta,s) = \Delta(\eta,-s)$ which can only be the case if the coefficients are rational functions in s^2.

If $b_0(s^2)$ is the least common denominator of the coefficients $\{a_i(s^2) \; ; \; i=1,2,\ldots,m\}$ then from (6.5.11)

$$b_0(s^2)\Delta(\eta,s) = b_0(s^2)\eta^m + b_1(s^2)\eta^{m-1} + \ldots + b_m(s^2) = 0$$

(6.5.13)

where the coefficients $\{b_i(s^2) \; : \; i = 1,2,\ldots,m\}$ are polynomials in s^2. The function of a complex variable $\eta(s)$, defined by (6.5.13) is an algebraic function. In general, (6.5.11) and (6.5.13) will be reducible into several irreducible equations over the field of rational functions in s, thereby defining a set of algebraic functions. However, for simplicity of exposition it will be assumed that equations (6.5.11) and (6.5.13) are irreducible over the field of rational functions in s.

If we compare equations (6.5.8) and (6.5.10) we see that the optimal closed-loop poles can be defined as the left half-plane solutions of

$$\eta(s) = -\rho$$ (6.5.14)

i.e. the optimal characteristic frequency loci are the $180°$ phase contours of the algebraic function $\eta(s)$ in the left half-plane. If we now consider ρ as a complex variable and substitute for $\eta(s)$ in (6.5.13) we obtain

$$b_0(s^2) \, (-\rho)^m + b_1(s^2) \, (-\rho)^{m-1} + \ldots + b_m(s^2) = 0$$

(6.5.15)

which is an algebraic equation defining the algebraic functions $s(\rho)$ and $\rho(s)$. The left half-plane branches of the algebraic function $s(\rho)$, for ρ real and positive, are the <u>optimal characteristic frequency loci</u>. (Note that equation (6.5.15) can be obtained directly from equation (6.5.8)).

To determine the asymptotic behaviour of the optimal characteristic frequency loci we need approximations to the branches of the algebraic function $s(\rho)$ in the neighbourhood of $s = \infty$, as ρ approaches zero. These can be obtained as the first terms in the series expansions for the branches of $s(\rho)$ about the point $s=\infty$, as ρ approaches zero. For this purpose we put $s = z^{-1}$ in equation (6.5.15) and procede as in section 6.2.

6.6   Example 2

To demonstrate the procedure the asymptotic behaviour of the optimal closed-loop poles will be calculated for

$$G(s) = \frac{1}{(s+1)(s+2)(s+3)(s+4)} \begin{bmatrix} s + 2 & 6 \\ s + 3 & 1 \end{bmatrix},$$

$$Q = I$$

and

$$\rho R = \rho I$$

From this data we have

$$H(s) \triangleq R^{-1}G^{T}(-s)QG(s)$$

$$= \frac{1}{(s^2-1)(s^2-2)(s^2-3)(s^2-4)} \begin{bmatrix} -2s^2 + 13 & -7s + 15 \\ 7s + 15 & 37 \end{bmatrix}$$

so that

$$\Delta(\eta,s) \triangleq \det\left[\eta I_2 - H(s)\right]$$

$$=\eta^2 - \frac{(-2s^2 + 50)}{(s^2-1)(s^2-2)(s^2-3)(s^2-4)}\eta + \frac{(-25s^2 + 256)}{(s^2-1)^2(s^2-2)^2(s^2-3)^2(s^2-4)^2}$$

$$= 0$$

If we now substitute $\eta = -\rho$, and multiply throughout by the least common denominator of the coefficients, we obtain

$$(s^2-1)^2(s^2-2)^2(s^2-3)^2(s^2-4)^2\rho^2+(s^2-1)(s^2-2)(s^2-3)(s^2-4)(-2s^2+50)\rho$$

$$+ (-25s^2 + 256) = 0.$$

The substitution $s = z^{-1}$ now gives

$$(1-z^2)^2(1-2z^2)^2(1-3z^2)^2(1-4z^2)^2\rho^2 + z^6(1-z^2)(1-2z^2)(1-3z^2)(1-4z^2)$$

$$(-2+50z^2)\rho$$

$$+ z^{14}(-25 + 256z^2) = 0$$

The Newton diagram for this last equation is shown in figure 34, from which we obtain

$\mu_1 = \frac{1}{6}$, $\mu_2 = \frac{1}{8}$, and hence the approximations $z \tilde{=} e_1\rho^{\frac{1}{6}}$ and

$z \approx e_2\rho^{\frac{1}{8}}$. To find the values of e_1 we substitute $z = e_1\rho^{\frac{1}{6}}$ into

$$\rho^2 - 2z^6\rho = 0 \quad \left\{ \begin{array}{l} \text{i.e. the terms corresponding to the} \\ \text{points on the first link of the Newton} \\ \text{diagram equated to zero,} \end{array} \right.$$

giving $e_1 = \sqrt[6]{0.5} = 0.891 \exp\left(\frac{j2k\pi}{6}\right)$, $k=0,1,2,3,4,5.$

If we substitute $z = e_2\rho^{\frac{1}{8}}$ into

$$-25z^{14} - 2z^6\rho = 0 \quad \left\{ \begin{array}{l} \text{i.e. the terms corresponding to the} \\ \text{points on the second link of the Newton} \\ \text{diagram equated to zero,} \end{array} \right.$$

we obtain

$$e_2 = \sqrt[8]{-0.08} = 0.729 \exp j\left(\frac{\pi+2k\pi}{8}\right), \quad k=0,1,2,\ldots,7$$

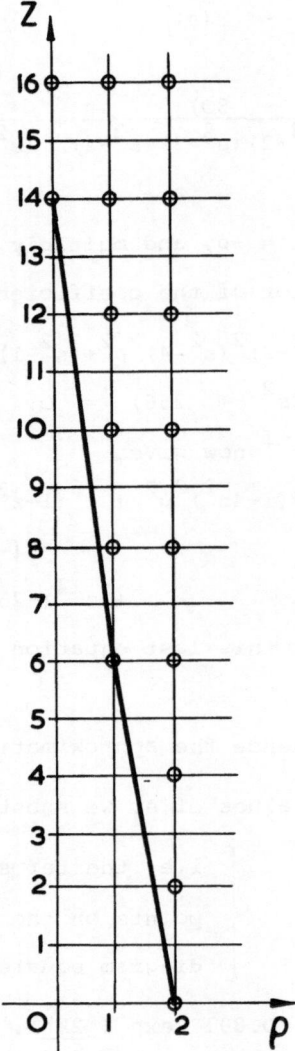

Figure 34. Newton diagram

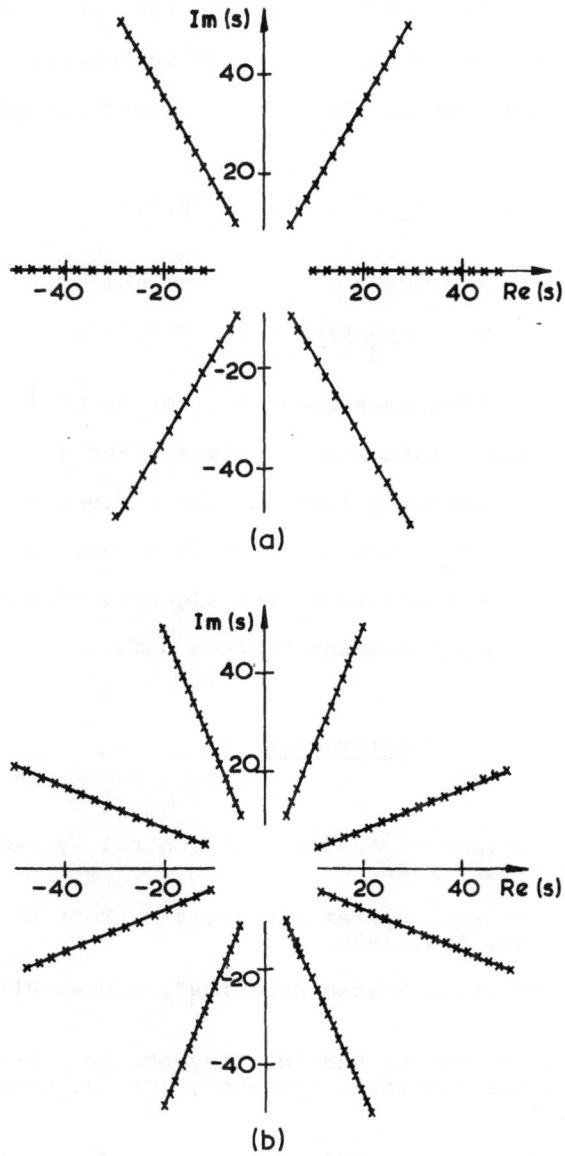

(a)

(b)

Figure 35. Asymptotic behaviour of the optimal characteristic
frequency loci (plus right half-plane image) displayed
on the Riemann surface domain of $\eta(s)$
(a) sheet 1 of the surface
(b) sheet 2 of the surface

If we now substitute back for s, and take only left half-plane solutions, we obtain the following expressions for the asymptotic behaviour of the optimal characteristic frequency loci

$$s \simeq 1.12 \left[\exp \left(j \frac{k\pi}{3} \right) \right] \rho^{-\frac{1}{6}}, k = 2,3,4$$

and

$$s \simeq 1.37 \left[\exp \ j \ \frac{(\pi+2k\pi)}{8} \right] \rho^{-\frac{1}{8}}, \ k = 2,3,4,5.$$

Note that using Kwakernaak's notation [11], these define Butterworth patterns of orders 3 and 4. The optimal characteristic frequency loci and their right half-plane image are shown in figure 35. Note that they have been computed as the 180° phase contours of the algebraic function $\eta(s)$ and displayed on its Riemann surface domain.

References

[1] W.R. Evans, "Graphical Analysis of Control Systems", Trans. AIEE, 67, 547-551, 1948.

[2] W.R. Evans, "Control System Synthesis by Root Locus Method", Trans. AIEE, 69, 1-4, 1950.

[3] W.R. Evans, "Control System Dynamics", McGraw-Hill, New York, 1954.

[4] B. Kouvaritakis, and U. Shaked, "Asymptotic behaviour of root-loci of multivariable systems", Int. J. Control, 23, 297-340, 1976.

[5] D.H. Owens, "A note on series expansions for multivariable root-loci", Int. J. Control, 26, 549-557, 1977.

[6] U. Shaked, "The angles of departure and approach of the root-loci in linear multivariable systems", Int. J. Control, 23, 445-457, 1976.

[7] G.A. Bliss, "Algebraic Functions", Dover, New York, 1966 (reprint of 1933 original).

[8]   B.A.Fuchs, and V.I.Levin, "Functions of a Complex Variable",
        International Series of Monographs in Pure and Applied
        Mathematics, Pergamon Press, 1961 (translation of 1951
        Russian original).

[9]   E.Hille, "Analytic Function Theory", Vol. 2, Ginn and Co.,
        U.S.A., 1962.

[10]   I.Postlethwaite, "A note on the characteristic frequency
        loci of multivariable linear optimal regulators", IEEE
        Trans. Automatic Control, 23, 757-760, 1978.

[11]   H. Kwakernaak, "Asymptotic Root Loci of Multivariable Linear
        Optimal Regulators", IEEE Trans. Automatic Control, 21,
        378-382, 1976.

[12]   A.G.J.MacFarlane, "Dual-system methods in dynamical analysis
        Pt. 2 - Optimal regulators and optimal servo-mechanisms",
        Proc. IEE, 1458-1462, 1969.

7. On parametric stability and future research

A feedback system is said to be stable if all of its closed-loop poles are in the left half-plane. The stability of a control system is therefore dependent on its associated parameters. Sometimes in a control system the value of a parameter is uncertain perhaps due to ageing, deterioration, or damage; in other instances it may be desirable, for economic reasons, to change a parameter value. In both these cases a technique which predicts the relative stability of a system with respect to a given parameter would be extremely useful.

A dominant theme in the preceding chapters has been the association of a system with two sets of algebraic functions: characteristic gain functions and characteristic frequency functions. In this chapter characteristic parameter functions are introduced, and used to develop the ideas of 'parametric' root loci and 'parametric' Nyquist loci from which the relative stability of a system, with respect to a single parameter, can be determined.

In the final section of this chapter a few tentative proposals and suggestions for future research are made.

7.1 Characteristic frequency and characteristic parameter functions

The feedback configuration considered is shown in figure 36, where $A(k_2,k_3,\ldots,k_q)$, $B(k_2,k_3,\ldots,k_q)$, $C(k_2,k_3,\ldots k_q)$ and $D(k_2,k_3,\ldots,k_q)$ are state-space matrices which are dependent on $(q-1)$ real, time-invariant parameters and k_1 is a scalar, time-invariant gain parameter common to all the loops.

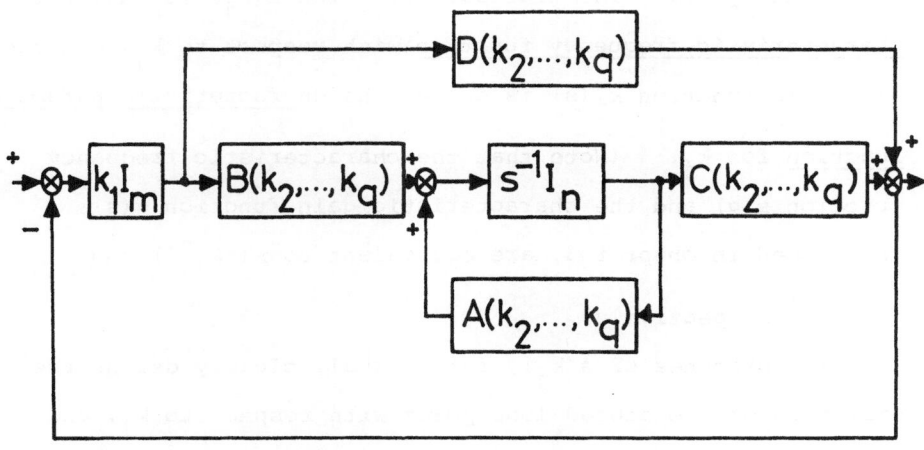

Figure 36. Feedback configuration
for parameter analysis

The closed-loop poles for this configuration are solutions of

$$\det \left[sI_n - S(k) \right] = 0 \qquad (7.1.1)$$

where

$$S(k) \triangleq A(k_2,\ldots,k_q) - B(k_2,\ldots,k_q)\left[k_1^{-1}I_m + D(k_2,\ldots,k_q)\right]^{-1} C(k_2,\ldots,k_q)$$

is the closed-loop frequency matrix [see section 3.1]. If numerical values for all the parameters except one, k_j say, are substituted into equation (7.1.1), and k_j considered as a complex variable, then the resulting algebraic equation (which for simplicity of exposition will be regarded as irreducible) defines a pair of algebraic functions [1], $s(k_j)$ and $k_j(s)$. The algebraic function $s(k_j)$ is called the <u>characteristic frequency function</u> with respect to k_j, and the algebraic function $k_j(s)$ is called the <u>characteristic parameter function</u> for k_j. (Note that the characteristic frequency function $s(g)$ and the characteristic gain function $g(s)$, introduced in chapter 3, are equivalent to $s(-k_1^{-1})$ and $-k_1(s)^{-1}$ respectively).

The branches of $s(k_j)$, for k_j real, clearly define the variation of the closed-loop poles with respect to k_j, and as such are termed <u>parametric root loci</u>. Alternatively, the parametric root loci can be viewed as the $0°$ phase contours of $k_j(s)$ on the frequency surface domain for $k_j(s)$.

Dual to the parametric root loci are the <u>parametric Nyquist loci</u> or <u>characteristic parameter loci</u> which are the branches of $k_j(s)$ as s traverses the imaginary axes. Alternatively, the characteristic parameter loci can be viewed as the $\pm 90°$ phase contours of $s(k_j)$ on the Riemann surface domain for $s(k_j)$ which will be called the <u>parameter surface</u> for k_j.

If, for a particular system, we have a set of nominal values for the system parameters we can determine which, if any, are sensitive with respect to stability by looking at the set of parameter surfaces. To help in such an assessment the following generalizations of gain and phase margin are introduced.

7.2   Gain and phase margins

The $\pm 90°$ phase contours of $s(k_j)$ on the parameter surface for k_j trace out the boundary between stable and unstable closed-loop poles and therefore we can define parameter gain and phase margins for k_j about a stable operating point $\overset{o}{k}_j$ which give a measure of the relative stability of the system with respect to k_j.

Parameter gain margin.   Parameter gain margin is defined with respect to a stable operating point $\overset{o}{k}_j$ as the smallest change in parameter gain about $\overset{o}{k}_j$ needed to drive the system into instability.   Let d_i be the shortest distance along the real axis from a stable operating point $\overset{o}{k}_j$ to the stability boundary (characteristic parameter loci) on the ith sheet of the parameter surface for k_j.   Then the parameter gain margin is defined as $\underset{i}{\min}\{d_i:\ i=1,2,\dots,n\}$.

Parameter phase margin.   On each of the n sheets of the parameter surface for k_j imagine that an arc is drawn, centre the origin, from a stable operating point $\overset{o}{k}_j$ until it reaches the stability boundary (characteristic parameter loci).   Let ϕ_i be the angle subtended at the origin by the corresponding arc on the ith sheet.   Then the parameter phase margin is defined as $\underset{i}{\min}\{\phi_i:\ i=1,2,\dots,n\}$

7.3   Example

In this section an inverted pendulum positioning system (see figure 37) is considered and its stability analysed with respect to one of its parameters, namely the mass of the carriage.

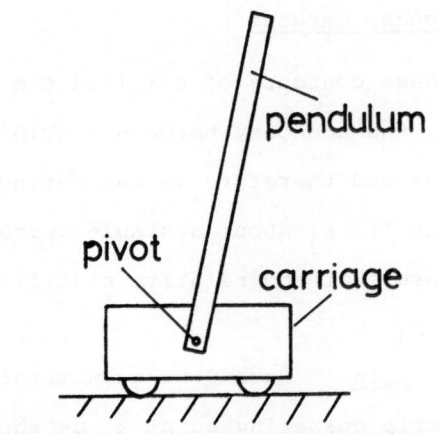

Figure 37. Inverted pendulum
positioning system

This system has also been used by Kwakernaak and Sivan [2], Cannon[3], and Elgerd[4]. The system can be modelled by the following linearized state differential equation[2]

$$x(t) = \begin{bmatrix} 0 & 1 & 0 & 0 \\ 0 & -\dfrac{F}{M} & 0 & 0 \\ 0 & 0 & 0 & 1 \\ -\dfrac{g}{L'} & 0 & \dfrac{g}{L'} & 0 \end{bmatrix} x(t) + \begin{bmatrix} 0 \\ \dfrac{1}{M} \\ 0 \\ 0 \end{bmatrix} u(t) \qquad (7.3.1)$$

where u(t) is a force exerted on the carriage by a small motor; M is the mass of the carriage; F is the friction coefficient associated with the movement of the carriage; and L' is given by

$$L' = \frac{J + mL^2}{mL} \qquad (7.3.2)$$

where m is the mass of the pendulum; L is the distance from
the pivot to the centre of gravity of the pendulum; and J
is the moment of inertia of the pendulum with respect to
the cente of gravity.

The system is stabilizable using state feedback of the
form

$$u(t) = -Kx(t) \tag{7.3.3}$$

and using the numerical values

$$\frac{F}{M} = 1 \text{ s}^{-1}$$

$$\frac{1}{M} = 1 \text{ kg}^{-1}$$

$$\frac{g}{L'} = 11.65 \text{ s}^{-2} \tag{7.3.4}$$

$$L' = 0.842m$$

it can be found [2] that

$$K = [86.81, \ 12.21, \ -118.4, \ -33.44] \tag{7.3.5}$$

stabilizes the linearized system placing the closed-loop
poles at $-4.706\pm j\,1.382$ and $-1.902\pm j3.420$.

We will now look at the parameter surface for M to see
how variations in the carriage mass, about an operating point
of 1kg, affect the stability of the system.   The four sheets
of the mass surface, characterized by constant phase and
magnitude contours of s(M), are shown in figures 38-41,
from which the following stability margins are obtained:

parameter (mass) gain margin = 1 kg

parameter (mass) phase margin = 60° .

The gain margin of 1kg corresponds to reducing the carr iage
mass to zero before instability occurs.   The parameter

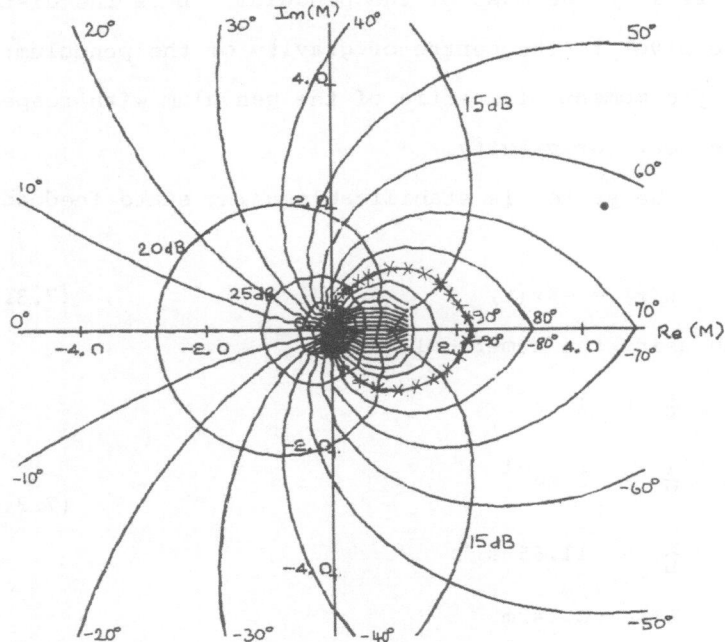

Figure 38. Sheet 1 of parameter (mass) surface

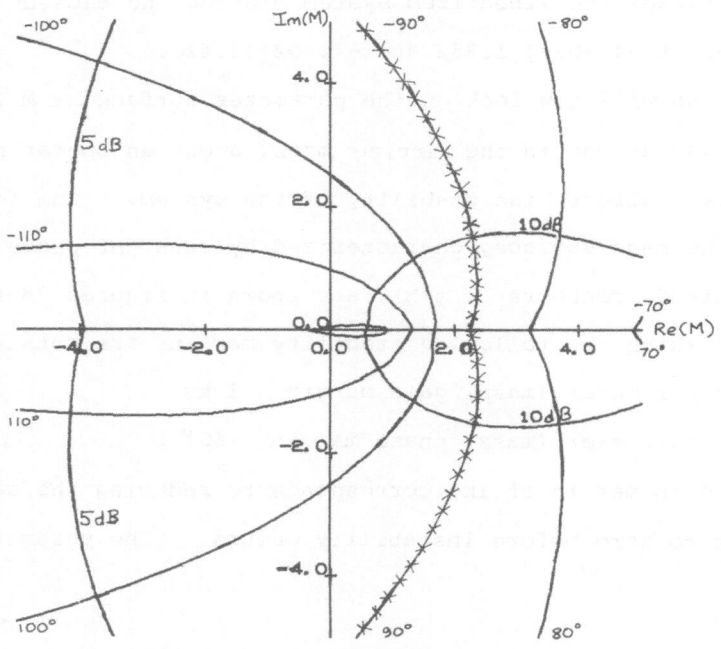

Figure 39. Sheet 2 of parameter (mass) surface

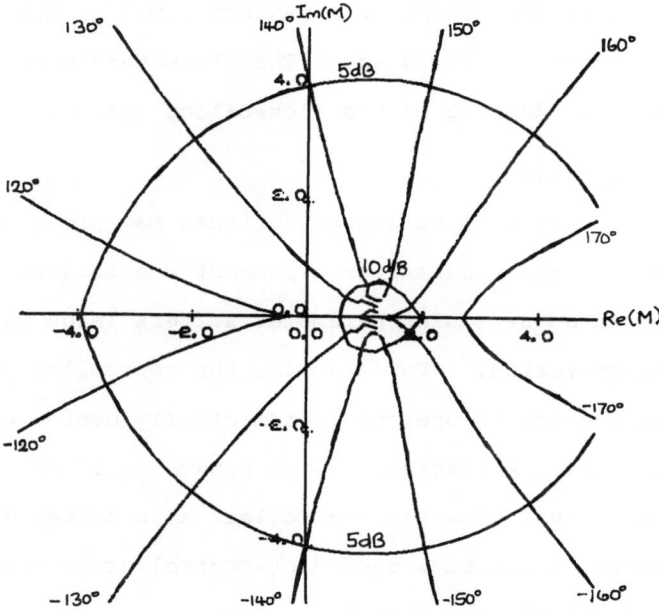

Figure 40. Sheet 3 of parameter (mass) surface

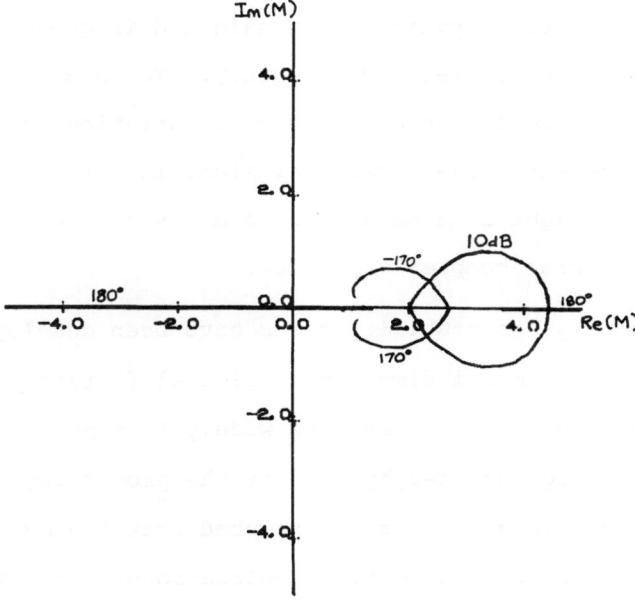

Figure 41. Sheet 4 of parameter (mass) surface

surface also shows that there is a maximum limit to the carriage
mass, for stability, of 2.125 kg.   The phase margin of 60°
indicates adequate damping of the closed-loop system.

7.4   Future research

It is thought that parameter surfaces may prove to be
useful in the design of parameter dependent controllers for
systems in which a particular parameter suffers large variations
during normal operation.   For example, the controller of an
aircraft engine needs to operate satisfactorily over a wide
range of altitudes.   A possible design scheme could be

(i)   to design real constant controllers at a number of altitudes,

(ii)   to obtain an altitude dependent controller by "matrix
         interpolation", and finally

(iii)   to analyse the stability of the system over the whole
         working range using the "altitude surface".

As indicated in this chapter the methods put forward in
this work are not only applicable to gain and frequency, but
any single system parameter and frequency.   To obtain stability
tests in terms of more than one parameter variation is a complicated
problem, but one with great practical significance.   It is felt
that valuable insight into this problem may stem from a study of
functions of several complex variables.

In recent years stability tests have been developed for
two-dimensional and multi-dimensional digital filters [5].   Two-
dimensional digital filters are used widely in many fields, such
as image processing, and geophysics for the processing of seismic,
gravity and magnetic data.   It is expected that higher
dimensional filters will also find applications;   for example,
three-dimensional filters in holography.   The stability tests
which have emerged are of a Nyquist-type and it is thought that

a deeper understanding of these developments by control engineers
may lead to suitable adaption for use in the control field;
possibly in the study of stability under multi-parameter
variations.

A constant theme throughout this work has been the
association of a multivariable system with one or possibly
more algebraic functions each of which is defined on an
appropriate Riemann surface. A Riemann surface for an
algebraic function is topologically equivalent to a sphere
with "handles", and the number of handles is known as the
genus number of the surface [6] . The genus number may prove
to be an important characteristic of a system and it is felt
that it may be related to "decoupling", that is, the transformation
of a multivariable system to a system which is effectively a set
of single-input, single-output systems. A single-input, single-
output system has a corresponding trivial (one sheeted) frequency
surface of genus zero, so that to decouple a multivariable
system would be equivalent to reducing an m-sheeted Riemann
surface, with a genus number greater than or equal to zero,
to a set of m trivial Riemann surfaces each of zero genus.
Consequently the establishment of relationships between branch-
point singularities (whose nature and location determine such
topological properties as the genus number) and decoupling
could be a fruitful line of research. In appendix 7 an
interesting relationship is developed which shows that a branch
point on a gain surface corresponds to a branch point or
stationary point on the corresponding frequency surface and
vice versa.

References

[1]   G.A.Bliss, "Algebraic Functions", Dover, New York,1966
        (reprint of 1933 original).

[2]   H. Kwakernaak, and R. Sivan, "Linear Optimal Control
        Systems", Wiley, New York, 1972.

[3]   R.H. Cannon, Jr, "Dynamics of Physical Systems",
        McGraw-Hill, New York, 1967.

[4]   O.I.Elgerd, "Control Systems Theory", McGraw-Hill,
        New York, 1967.

[5]   E. I. Jury, "Inners and Stability of Dynamical Systems",
        Wiley, New York, 1974.

[6]   G. Springer, "Introduction to Riemann Surfaces",
        Addison-Wesley, Reading, Mass., 1957.

Appendix 1. Definition of an algebraic function

Let $\Lambda(q,v)$ be a polynomial in q of the form

$$\Lambda(q,v) = f_o(v)q^m + f_1(v)q^{m-1}+\ldots+f_m(v) \qquad (A1.1)$$

where each coefficient $\{f_i(v): i=1,2,\ldots,m\}$ is itself a polynomial in v with coefficients in the domain of complex numbers.   Then an algebraic function is a function q(v) defined for values of v in the complex v- plane by an equation of the form

$$\Lambda(q,v) = 0 \qquad\qquad (A1.2)$$

The polynomial $\Lambda(q,v)$ can be rewritten as a polynomial in v with coefficients which are themselves polynomials in q, and when considered in this way equation (A1.2) defines an algebraic function v(q).

For a fixed value of v, v_o say, equation (A1.2) has m solutions which are called branches of q(v), and in the neighbourhood of v_o the branches are representable by power series expansions [1].

It is assumed in the above definition that $\Lambda(q,v)$ is an irreducible polynomial in (q,v), that is, that $\Lambda(q,v)$ is not the product of two or more polynomials in (q,v).   If $\Lambda(q,v)$ was expressible as a product of t polynomials in (q,v), equation (A1.2) would then define t algebraic functions of the form q(v), or t algebraic functions of the form v(q).

Appendix 2.   A reduction to the irreducible rational canonical form

In this appendix a proposed method is given for reducing any m-square matrix to its <u>irreducible</u> rational canonical form.   The method used is a variation on that given

by Ayres $[2]$ for finding the rational canonical form of a
square matrix.   The rational canonical form of a square
matrix G is similar to the irreducible form except that its
diagonal blocks correspond to the invariant factors of gI-G,
rather than the irreducible factors.   The procedure given
by Ayres $[2]$ for finding the rational canonical form is
outlined below along with some necessary definitions.

Definitions: If the vectors

$$X, GX, G^2 X, \ldots, G^{t-1} X \qquad (A2.1)$$

are linearly independent but

$$X, GX, G^2 X, \ldots, G^{t-1} X, G^t X \qquad (A2.2)$$

are not then (A2.1) is called a chain of length t having X
as its leader.

Procedure: For a given m-square matrix G over any field F :

(i)   let X_m be the leader of a chain C_m of maximum length
for all m-vectors over F ;

(ii)   let X_{m-1} be the leader of a chain C_{m-1} of maximum
length (any member of which is linearly independent of the
preceding members and those of C_m) for all m-vectors over F
which are linearly independent of the vectors of C_m ;

(iii) let X_{m-2} be the leader of a chain C_{m-2} of maximum
length (any member of which is linearly independent of the
preceding members and those of C_m and C_{m-1}) for all m-
vectors over F which are linearly independent of the vectors
of C_m and C_{m-1} ;

and so on.   Then, for

$$E = \left[X_j, GX_j, \ldots G^{t_j-1} X_j; X_{j+1}, GX_{j+1}, \ldots, G^{t_{j+1}-1} X_{j+1}; \right.$$

$$\left. \ldots ; X_m, GX_m, \ldots, G^{t_m-1} X_m \right]$$

we have that $E^{-1}GE$ is the rational canonical form of G.

In this approach the chains of maximum length are used to pick out the invariant factors. Now the invariant factors are made up from products of the irreducible characteristic equations which are required in the irreducible rational canonical form. Therefore, if instead of using chains of maximum length those of minimum length are found, the transformation matrix E so formed is that required to give the irreducible rational canonical form Q.

The problem with both of these methods is that no indication is given which enables one to know when a chain of maximum or minimum length has been obtained, except that in the contrary case there will appear a chain of longer or shorter length.

It is interesting to note that the chains of maximum and minimum length form bases for the G-invariant subspaces of maximum and minimum dimensions respectively. Therefore the problem of finding the chains of minimum length is equivalent to that of finding the invariant suhspaces of minimum dimension. It also happens that just as the 1-dimensional invariant subspace (eigenvector) picks out an eigenvalue, a t-dimensional invariant subspace (from the set of those of minimum dimensions) picks out an irreducible equation of degree t, defining t eigenvalues.

Example: To find the irreducible rational canonical form of

$$G(s) = \begin{bmatrix} \dfrac{(s+3)}{(s+1)^2} & \dfrac{(s+2)}{(s+1)^2} & 0 \\[3mm] \dfrac{(s-3)}{(s+1)^2} & \dfrac{-2}{(s+1)^2} & 0 \\[3mm] \dfrac{-1}{(s+1)^2} & \dfrac{-(s+2)}{2(s+1)^2} & \dfrac{1}{s+1} \end{bmatrix}$$

Let X = $(0\ 0\ 1)^t$, where "t" denotes the transpose,

then $GX = \dfrac{1}{(s+1)}\ X$

and X is a chain of minimum length.

Let $Y = (0\ 1\ 0)^t$

then $GY = \left[\dfrac{s+2}{(s+1)^2}\quad \dfrac{-2}{(s+1)^2}\quad \dfrac{-(s+2)}{2(s+1)^2} \right]^t$

and $G^2Y = \left[\dfrac{s+2}{(s+1)^3}\quad \dfrac{s-2}{(s+1)^3}\quad \dfrac{-(s+2)}{2(s+1)^3} \right]^t = \dfrac{1}{(s+1)}\ GY + \dfrac{s}{(s+1)^3}\ Y$

No chain of smaller length can be found and therefore Y,GY completes the set of minimum length chains. The transformation matrix E(s) is therefore given by

$$E(s) = \left[X\quad Y\quad GY \right] = \begin{bmatrix} 0 & 0 & \dfrac{s+2}{(s+1)^2} \\[2ex] 0 & 1 & \dfrac{-2}{(s+1)^2} \\[2ex] 1 & 0 & \dfrac{-(s+2)}{2(s+1)^2} \end{bmatrix}$$

and

$$Q(s) = E^{-1}(s)G(s)E(s)$$
$$= \begin{bmatrix} \dfrac{1}{s+1} & 0 & 0 \\[2ex] 0 & 0 & \dfrac{s}{(s+1)^3} \\[2ex] 0 & 1 & \dfrac{1}{(s+1)} \end{bmatrix}$$

which implies that the irreducible characteristic equations are

$$\Delta_1(g,s) = g - \dfrac{1}{s+1}$$

$$\Delta_2(g,s) = g^2 - \dfrac{1}{(s+1)}\ g - \dfrac{s}{(s+1)^3}$$

Note that these are also obvious from the dependence relations obtained when finding the chains of minimum length.

Appendix 3. The discriminant

In this appendix two methods are given for finding
the discriminant of an equation of the form

$$\Phi(g,s) = b_o(s)g^t + b_1(s)g^{t-1} + \ldots + b_t(s) = 0$$

Method 1 (Barnett [3])

The resultant $R[a(g), c(g)]$ of two polynomials
$a(g)$ and $c(g)$, given by

$$a(g) = a_o g^n + a_1 g^{n-1} + \ldots + a_n$$

$$c(g) = c_o g^m + c_1 g^{m-1} + \ldots + c_m$$

where $a_i, c_i \in \mathbb{C}$, is the determinant

$$R[a(g), c(g)] = \begin{vmatrix} a_o & a_1 & a_2 & \cdots & & a_n & 0 & \\ & 0 & a_o & a_1 & \cdots & & a_{n-1} & a_n \\ & \cdot & \cdot & \cdot & \cdots & & & \\ & \cdot & \cdot & \cdot & \cdots & & & a_n-1 & a_n \\ & \cdot & \cdot & \cdot & \cdots & c_o & c_1 & \cdot\cdot & c_{m-1} & c_m \\ & \cdot & \cdot & \cdot & \cdots c_o & c_1 & c_2 & \cdot\cdot & c_m & 0 \\ & \cdot & \cdot & \cdot & \cdots & & & \\ & c_o & c_1 & c_2 & \cdots & & & \end{vmatrix} \begin{matrix} \\ \\ \\ m \\ \text{rows} \\ \\ n \\ \text{rows} \end{matrix}$$

The polynomials $a(g)$ and $c(g)$ have a common factor (of
degree greater than zero) if and only if the $(n+m)$ -order
determinant $R[a(g), c(g)]$ is zero, provided that a_o
and c_o are not both zero.

Let the derivative with respect to g of the polynomial
$a(g)$ be denoted by $a'(g)$. Then the discriminant of the
polymonial $a(g)$ is the determinant $D_g(a_o, a_1, \ldots, a_n)$ defined

by

$$D_g(a_o, \ldots, a_n) = \frac{1}{a_o} \ R\left[\ a(g), a'(g)\ \right]$$

The polynomial $a(g)$ has a repeated factor if and only if
the discriminant $D_g(a_o, \ldots, a_n)$ is zero.

Now consider a polynomial of the type

$$\Phi(g,s) = b_o(s)g^t + b_1(s)g^{t-1} + \ldots + b_t(s) , \quad t > 0$$

where the coefficients $\{b_i(s) : i = 1, 2, \ldots, t\}$ are all

polynomials in s. The discriminant of this polynomial in
g is found from $D_g(b_o, b_1, \ldots, b_t)$ as defined above by replacing
$b_o, \ldots, b_t$ by $b_o(s), \ldots, b_t(s)$ respectively. Thus there is a
function $D_g(s)$ again called the discriminant and defined by

$$D_g(s) = \frac{1}{b_o(s)} \ R\left[\ \Phi(g,s), \Phi'(g,s)\ \right]$$

where $\Phi'(g,s)$ is the derivative with respect to g
of $\Phi(g,s)$.

Method 2   (Sansone and Gerretsen [4])

Consider the polynomial

$$a(g) = a_o g^n + a_1 g^{n-1} + \ldots + a_n , \quad n > 0$$

where $a_i \in \mathbb{C}$; then the discriminant of $a(g)$ is given by
the expression

$$D_g(a_o, \ldots, a_n) = a_o^{2n-2} P$$

where P is a determinant given by

$$P = \begin{vmatrix} \sigma_o & \sigma_1 & \cdots & \sigma_{n-1} \\ \sigma_1 & \sigma_2 & \cdots & \sigma_n \\ \cdot & \cdot & \cdot & \cdot \\ \sigma_{n-1} & \sigma_n & \cdots & \sigma_{2n-2} \end{vmatrix}$$

and the elements $\{\sigma_i : i = 1, 2, \ldots, 2n-2\}$ are functions of the coefficients $\{a_i : i = 0,1,\ldots,n\}$, and

$$\sigma_o = n$$

The elements $\sigma_1,\ldots,\sigma_{n-1}$ can be found from

$$a_1 + a_o\sigma_1 = 0$$

$$2a_2 + a_1\sigma_1 + a_o\sigma_2 = 0$$

$$\cdots\cdots\cdots\cdots\cdots$$

$$(n-1)a_{n-1} + a_{n-2}\sigma_1 + \cdots + a_o\sigma_{n-1} = 0$$

and $\sigma_n,\ldots,\sigma_{2n-2}$ from

$$a_n\sigma_o + a_{n-1}\sigma_1 + \cdots + a_o\sigma_n = 0$$

$$a_n\sigma_1 + a_{n-1}\sigma_2 + \cdots + a_o\sigma_{n+1} = 0$$

$$\cdots\cdots\cdots\cdots\cdots\cdots$$

$$a_n\sigma_m + a_{n-1}\sigma_{m+1} + \cdots + a_o\sigma_{n+m} = 0$$

$$\cdots\cdots\cdots\cdots\cdots\cdots$$

Consider now the polynomial $\Phi(g,s)$ given by

$$\Phi(g,s) = b_o(s)g^t + b_1(s)g^{t-1} + \cdots + b_t(s)$$

The discriminant of this polynomial in g is found from $D_g(b_o,\ldots,b_t)$ as defined above by replacing $b_o,\ldots,b_t$ by $b_o(s),\ldots,b_t(s)$ respectively. The polynomial $\Phi(g,s)$ therefore has repeated factors if and only if

$$D_g(s) = b_o^{2t-2}(s) P$$

is zero, where P is the determinant of a matrix whose elements are functions of the coefficients $\{b_i(s) : i = 0,1,2,\ldots,t\}$.

Appendix 4. A method for constructing the Riemann surface
domains of the algebraic functions corresponding
to an open-loop gain matrix G(s)

In sub-section 3.3-4 it was shown that for a characteristic
gain function of degree t the corresponding Riemann surface
is made up from t sheets of the complex s-plane stitched
together along cuts made between branch points and infinity.
Although the cuts are in some sense arbitrary (i.e. there is
no unique set of cuts), it is still a problem to choose a
set that is consistent, and then to be able to identify in what
order the sheets are connected together. In this appendix
a systematic method is given for solving this problem. The
method is quite elegant in that the resulting cuts are symmetrical
about the real axis and are always parallel to the imaginary
axis except for possible cuts along the real axis. Also,
the approach uses the open-loop gain matrix directly, so
that there is no need to find the characteristic equations,
and the resulting non-connected sets of connected sheets
represent the Riemann surfaces for the irreducible characteristic
equations.

The method is based on finding the eigenvalues of the
matrix for a grid of values covering the s-plane and then
sorting the eigenvalues in a continuous form along certain
lines of the grid. If the transfer function is of order m×m,
m arrays which will effectively represent the m sheets of
the Riemann surfaces are needed to store the eigenvalues.
The process is analogous to the analytic continuation
procedure described in sub-section 3.3-4 where individual

points of a circular disc are made bearers of unique functional values. Here individual points of the s-plane sheets (arrays) are being made the bearers of functional values.

The first line along which the eigenvalues are calculated and sorted is the real axis, and then the calculation and sorting is carried out along lines parallel to the imaginary axis and emanating from the real axis, as shown in figure 42. The calculation is only necessary in the upper half-plane since the eigenvalues in the lower half-plane are the complex conjugate of those in the upper half. Continuing this process the s-plane is covered, and m arrays of eigenvalues are obtained which are continuous along the real axis and along lines parallel to the imaginary axis but not necessarily crossing the real axis.

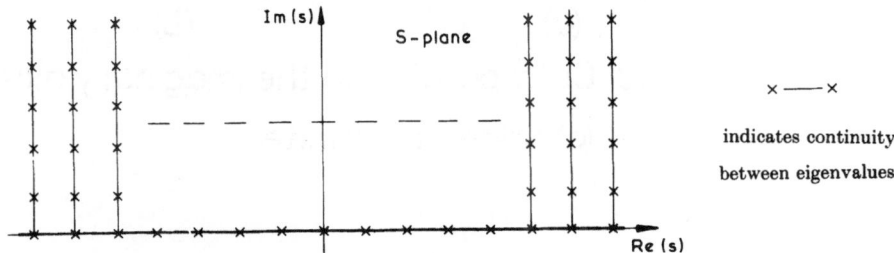

Figure 42. Lines along which eigenvalues
are calculated and sorted

By observation of each array or sheet the necessary cuts are obvious. If parallel to the imaginary axis a line of eigenvalues is not continuous with an adjacent line then these must be separated by a cut starting at a branch point and ending at a branch point or infinity as shown in figures 43 (a) and (b). Eigenvalues corresponding to values of s in the upper half-plane are complex conjugate to those in the lower half-plane and therefore continuity of eigenvalues across the real axis is impossible if the eigenvalues situated on the real axis are complex. Therefore a cut along the real axis is necessary whenever the real axis eigenvalues are complex. Again all the cuts start at a branch point and end at a branch point or infinity, as shown in figure 44.

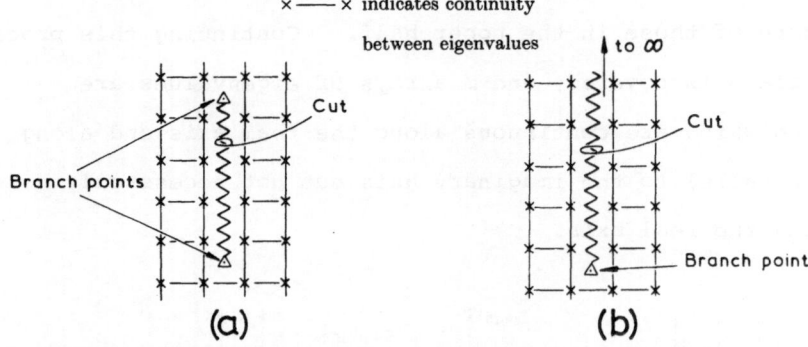

Figure 43. Cuts parallel to the imaginary axis
(a) finite (b) infinite

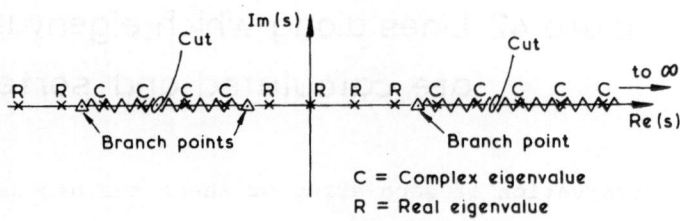

Figure 44. Cuts along the real axis

The stitching together of the sheets also becomes obvious by matching eigenvalues on one side of a cut on one sheet to those on another. When the matching process is complete, in general, there will be sets of connected sheets each set defining a Riemann surface.

To facilitate the identification of cuts and the matching of sheets it is very useful to draw the constant phase and constant magnitude contours on each sheet; this is demonstrated in the examples given in the main text.

Computationally sorting the eigenvalues along the real axis can be difficult because of the likelihood of real axis poles. This problem is overcome if the first line of calculation and sorting is changed to be a line parallel to, and a "large" distance away from, the real axis. The rest of the calculations and sorting are then carried out along lines parallel to the imaginary axis starting at this new line and finishing at the real axis; as shown in figure 45. With this new approach the only possible cuts are either along the real axis, as before (see figure 44) or between complex conjugate branch points as shown in figure 46.

A Riemann surface construction for

$$ G(s) = \frac{1}{1.25(s+1)(s+2)} \begin{bmatrix} 5s-2 & 2s-1 \\ 3s-18 & s-8 \end{bmatrix} $$

is shown using the original method in figures 47 and 48, and using the modified method in figures 49 and 50, and illustrates the arbitrariness of the cuts.

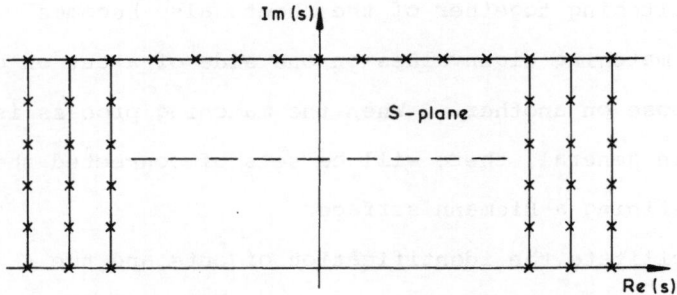

Figure 45. Lines along which eigenvalues are calculated and sorted (modified method)

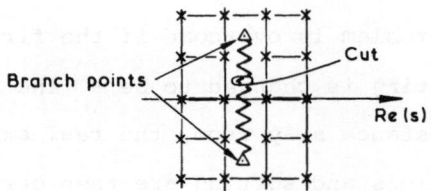

Figure 46. Cut parallel to the imaginary axis (modified method)

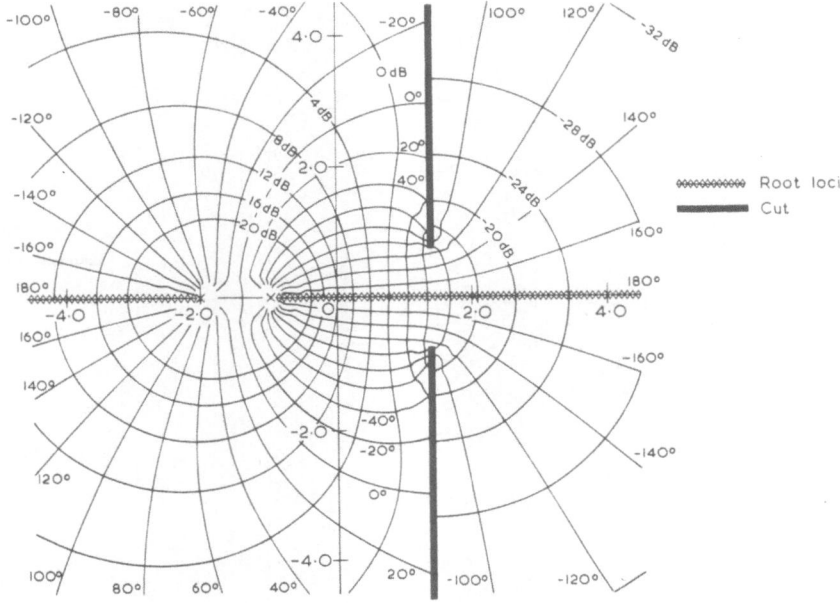

Figure 47. Sheet 1 of the Riemann surface

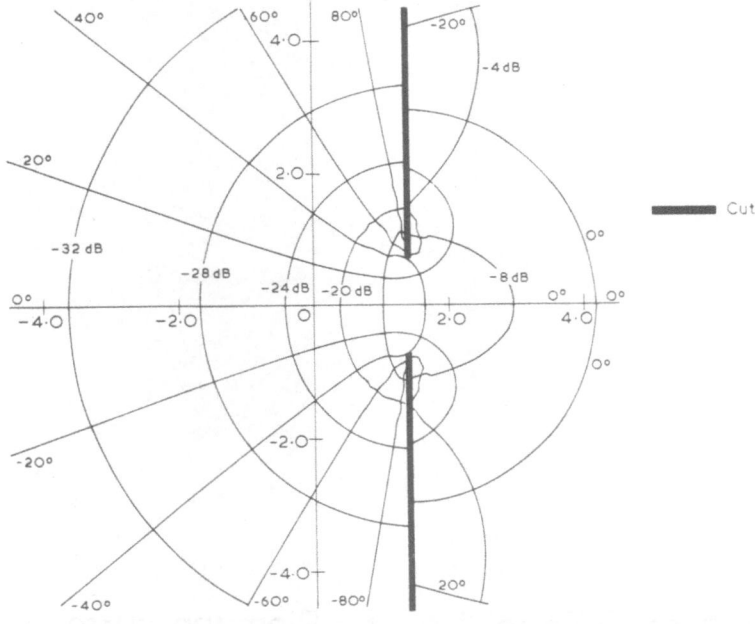

Figure 48. Sheet 2 of the Riemann surface

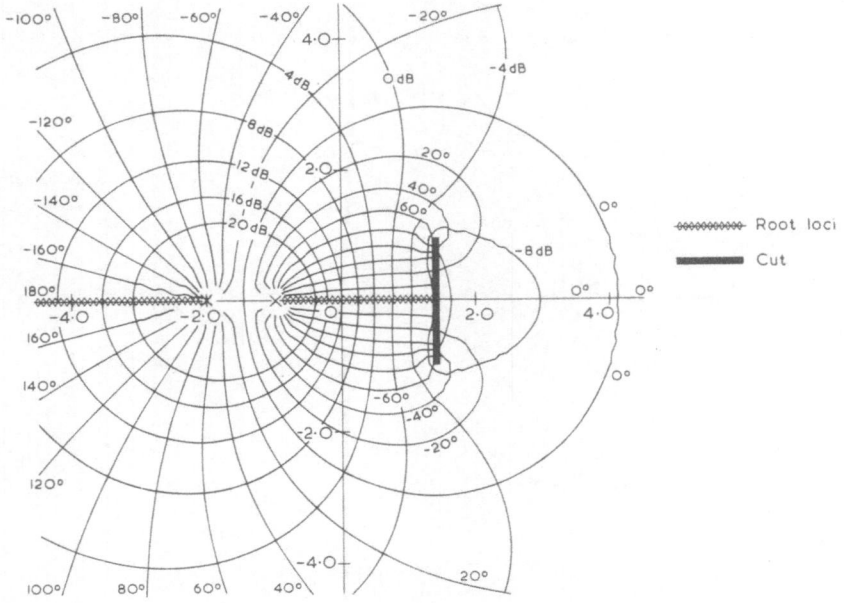

Figure 49. Sheet 1 of the Riemann surface
(modified method of construction)

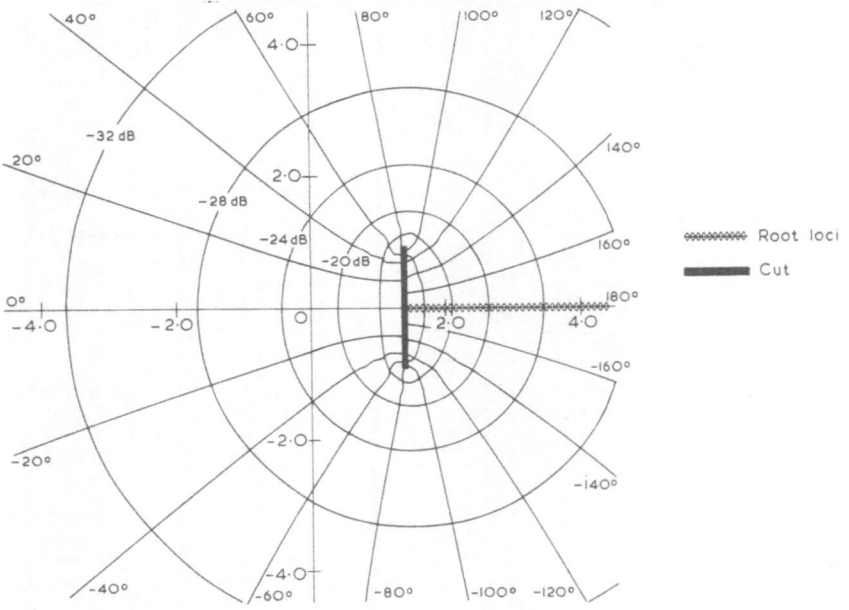

Figure 50. Sheet 2 of the Riemann surface
(modified method of construction)

Appendix 5.   Extended Principle of the Argument

The required extension of the Argument Principle does
not seem to be readily available in the literature, although
a suitable statement of the Principle for a general multiple-
valued analytic function has recently appeared in a text by
Evgrafov [5, page 98].   Therefore, to justify its use in
chapters 4 and 5, an appropriately extended Principle of the
Argument is developed here.

A5.1   Introduction

The extension required is non-trivial, with two main
problems to be overcome. The first of these arises from
the fact that, in general, the Riemann surface of an algebraic
function will be multiply connected.   For an example of this
source of difficulty consider a characteristic gain function
$g(s)$ associated with a 2×2 open-loop gain matrix G(s).
Suppose that $g(s)$ has four branch points, and that each
branch point is associated with a cycle of two sheets [6];
then the genus number [6; 7] of the associated Riemann surface
is one.   Further suppose that these branch points are disposed
in the frequency plane (s-plane) in the way shown in figure 51.
Then, taking two copies of the complex number sphere (Riemann
number sphere), making cuts between the branch points, and
forming the topological equivalent of the Riemann surface
in the usual way [7] one obtains a torus, as illustrated in
figures 52 and 53.   The region Ω, shown shaded on this torus
and having a boundary $\partial\Omega$, corresponds to the interiors of the
pair of Nyquist D-contours (shown in figure 52) for the
original complex number spheres out of which the torus was
constructed.   To cope with such a situation we must ensure
that the extended version of the Argument Principle holds for

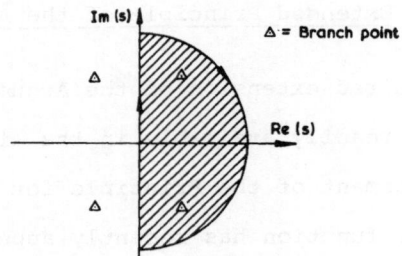

Figure 51. Frequency plane

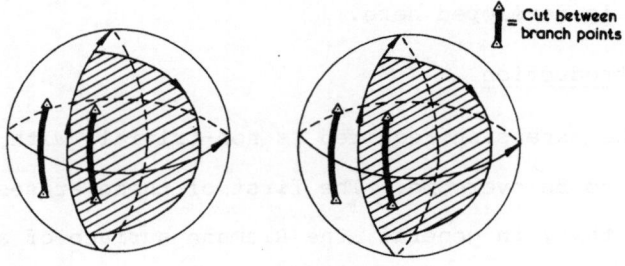

Figure 52. Two copies of the complex number sphere

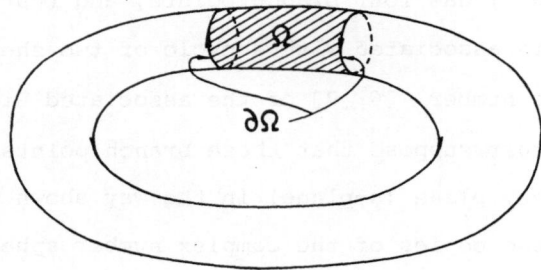

Figure 53. Riemann surface shown topologically
equivalent to a torus

a region of a Riemann surface which is non-simply connected
and whose boundary $\partial\Omega$ consists of several distinct closed
Jordan contours on the surface.   This basic extension of the
principle is achieved via a suitable generalization of the Cauchy
Residue Theorem and is covered in section A5.2.

The second obstacle to be overcome in a derivation of a
suitably extended Argument Principle is more directly
associated with the branch points of an algebraic function; in
this case with their effect on the calculation of residues
via contour integrals taken round the boundary $\partial\Omega$ of a region
Ω having branch points in its interior.   This problem is
overcome by a change of variable and is treated in section A5.3.

Finally, in section A5.4 the results of sections A5.2 and
.3 are combined to give the required extended Principle of the
Argument.

A5.2   Generalized form of Cauchy's Residue Theorem

The basic requirement is a generalization of Cauchy's
Residue Theorem to complex functions defined on the Riemann
surface of some known algebraic function, where the region
of integration may be non-simply connected and have a boundary
consisting of one or more closed Jordan contours.   Such a
generalization may be found in Bliss $\begin{bmatrix} 6 \end{bmatrix}$; the main theorem
is given below with slight rephrasing and change of notation
for the context of this work .   Two observations may be useful
in helping the reader unfamiliar with Riemann surface theory to
understand this theorem.

(i)   Each non-singular place (or point) on the Riemann
surface of an algebraic function $f(s)$ is uniquely defined by a

pair of values (f,s) which satisfy a known algebraic equation

$$\Delta(f,s) = 0$$

Thus at each point on the Riemann surface the value of a complex function $\Psi(f,s)$ of the <u>pair</u> of complex variables s and f is defined.

(ii) The Riemann surface for an algebraic function is an orientable surface.   Thus one can unambiguously define a positive sense in which a boundary $\partial\Omega$ can be traversed so as to "enclose" some region Ω .   The simplest way in which to visualize this procedure is to imagine someone walking round the boundary $\partial\Omega$ on a two-dimensional manifold in which Ω lies (embedded in a familiar space of three dimensions). The positive sense of traversal may then be taken as that in which someone walking <u>forward</u> round any portion of $\partial\Omega$ will always have Ω on his <u>right-hand</u> side.   In figure 53 the boundary Ω is shown with a positive orientation.

Generalized Residue Theorem

Let $\mathbb{R}$ be the Riemann surface of an algebraic function f(s) and Ω a portion of $\mathbb{R}$ not necessarily simply connected but having a boundary $\partial\Omega$ consisting of one or more closed Jordan contours not passing through any singular place on $\mathbb{R}$. Then if a function $\Psi(f,s)$ of the places on $\mathbb{R}$ is analytic on Ω and $\partial\Omega$ except possibly for a finite number of singular points in Ω, we have

$$\frac{1}{2\pi j} \int_{\partial\Omega} \Psi(f,s)\,ds = \Sigma \text{ residues of } \Psi(f,s) \text{ in } \Omega$$

$$(A5.2.1)$$

where the boundary $\partial\Omega$ is traversed in the positive sense with respect to Ω.

To derive the extended Argument Principle we will first apply the Generalized Residue Theorem to the function f'(s)/f(s), where f'(s) denotes the derivative of f(s) with respect to s, to give

$$\frac{1}{2\pi j} \int_{\partial\Omega} \frac{f'(s)}{f(s)} \, ds \; = \; \Sigma \text{ residues of } \frac{f'(s)}{f(s)} \text{ in } \Omega$$

Note that f(s) must have no poles or branch points (either of which are referred to as singular places) on the boundary $\partial\Omega$. For the purposes of deriving an Argument Principle we will now also exclude any zeros of f(s) from $\partial\Omega$. The next step is to calculate the residues of f'(s)/f(s) in Ω.

A5.3 Calculation of Residues

The poles, zeros and branch points of f(s) are singularities of $\frac{f'(s)}{f(s)}$, a fact which will become clearer as this section progresses. We will consider all the branch points of f(s) but it turns out that only those which are also poles or zeros of f(s) are relevant.

In the neighbourhood of a pole or a zero (not including those which are associated with a branch point) f(s) can be represented by a series which defines a single-valued function. For a pole p_i of order t_{p_i} the algebraic function can be represented by

$$f(s) \; = \; \sum_{n=-t_{p_i}}^{\infty} a_n (s-p_i)^n, \quad a_{-t_{p_i}} \neq 0 \qquad (A5.3.1)$$

i.e. $$f(s) \; = \; \frac{\phi(s)}{(s-p_i)^{t_{p_i}}} \qquad (A5.3.2)$$

where $\phi(p_i) \neq \infty$ and $\phi(s)$ is analytic in the neighbourhood of p_i. This gives

$$\frac{f'(s)}{f(s)} = \frac{-t_{p_i}}{(s-p_i)} + \frac{\phi'(s)}{\phi(s)} \qquad (A5.3.3)$$

and since $\frac{\phi'(s)}{\phi(s)}$ is analytic in the neighbourhood of p_i

the function $\frac{f'(s)}{f(s)}$ has a simple pole of residue $-t_{p_i}$ at

$s = p_i$.

For a zero s_i of order t_{z_i} the algebraic function can be represented by the series

$$f(s) = \sum_{n=t_{z_i}}^{\infty} c_n (s-z_i)^n, \; c_{t_{z_i}} \neq 0 \qquad (A5.3.4)$$

i.e. $f(s) = (s - z_i)^{t_{z_i}} \theta(s)$ \qquad (A5.3.5)

where $\theta(z_i) \neq 0$ and $\theta(s)$ is analytic in the neighbourhood of z_i. This gives

$$\frac{f'(s)}{f(s)} = \frac{t_{z_i}}{(s - z_i)} + \frac{\theta'(s)}{\theta(s)} \qquad (A5.3.6)$$

and we see that $\frac{f'(s)}{f(s)}$ has a simple pole of residue

t_{z_i} at $s = z_i$.

To determine the residues at a branch point it is necessary to make a change of variable [6, page 81].   The procedure is as follows.

In the neighbourhood of a finite branch point b_i

the algebraic function can be represented by a series of the form [8],

$$f(s) = \sum_{n=-\infty}^{\infty} d_n (\sqrt[r]{s - b_i})^n \qquad (A5.3.7)$$

where r is the number of sheets which form a cycle at the branch point. If the branch point is a pole of order $t_{p_i}^b$ and consists of a cycle of r sheets we have the series expansion

$$f(s) = \sum_{n=-t_{p_i}^b}^{\infty} d_n (\sqrt[r]{s-b_i})^n, \quad d_{-t_{p_i}^b} \neq 0 \qquad (A5.3.8)$$

Similarly if the branch point is a zero of order $t_{z_i}^b$ we have that

$$f(s) = \sum_{n=t_{z_i}^b}^{\infty} d_n (\sqrt[r]{s-b_i})^n, \quad d_{t_{z_i}^b} \neq 0 \qquad (A5.3.9)$$

otherwise

$$f(s) = \sum_{n=0}^{\infty} d_n (\sqrt[r]{s-b_i})^n, \quad d_o \neq 0 \qquad (A5.3.10)$$

These expansions are multivalued and therefore not suitable for determining residues. If, however, we make the substitution

$$s = b_i + x^r \qquad (A5.3.11)$$

we find that the series expansions (A5.3.8, 9 and 10) become

$$f(x) = \sum_{n=-t_{p_i}^b}^{\infty} d_n x^n, \quad d_{-t_{p_i}^b} \neq 0 \qquad (A5.3.12)$$

$$f(x) = \sum_{n=t_{z_i}^b}^{\infty} d_n x^n, \quad d_{t_{z_i}^b} \neq 0 \qquad (A5.3.13)$$

and

$$f(x) = \sum_{n=0}^{\infty} d_n x^n, \quad d_0 \neq 0 \qquad \qquad (A5.3.14)$$

respectively, where it is to be understood that

$$f(x) \equiv f(b_i + x^r) \qquad \qquad (A5.3.15)$$

The substitution has therefore mapped the algebraic function onto the x-plane where it can be represented by a "single-valued" series expansion.

By definition $[6]$ the residue at b_i of $\frac{f'(s)}{f(s)}$ analytic in a neighbourhood of b_i except possibly at b_i itself is

$$(\text{residue})_{b_i} \triangleq \frac{1}{2\pi j} \int_C \frac{f'(s)}{f(s)} \, ds \qquad \qquad (A5.3.16)$$

where C is a closed positively oriented Jordan contour on $\mathbb{R}$ bounding a neighbourhood N of b_i in which $\frac{f'(s)}{f(s)}$ is analytic except possibly at b_i. If we substitute for s from (A5.3.11) we find that

$$(\text{residue})_{b_i} \triangleq \frac{1}{2\pi j} \int_C \frac{f'(s)}{f(s)} \, ds$$

$$= \frac{1}{2\pi j} \int_{C_x} \frac{f'(x)}{f(x)} \frac{dx}{ds} \, ds$$

$$= \frac{1}{2\pi j} \int_{C_x} \frac{f'(x)}{f(x)} \, dx \qquad \qquad (A5.3.17)$$

where C_x is a closed positively oriented Jordan contour on the x-plane bounding a neighbourhood of the origin. The residue of $\frac{f'(s)}{f(s)}$ at a branch point b_i is therefore equal to the residue of $\frac{f'(x)}{f(x)}$ at the origin. The residue of the function

$\frac{f'(x)}{f(x)}$ is easily obtained from its series expansion . Following

the same procedure as before we note that $\frac{f'(x)}{f(x)}$ in the

neighbourhood of the origin has:

  (i) a simple pole with residue $-t^b_{p_i}$, if the branch

       point is a pole of order $t^b_{p_i}$; or

 (ii) a simple pole with residue $t^b_{z_i}$, if the branch

       point is a zero of order $t^b_{z_i}$;

otherwise $\frac{f'(x)}{f(x)}$ is analytic.

A5.4   Extended Principle of the Argument.

Combining the results of section A5.3 with that of

equation (A5.2.2) we have that

$$\frac{1}{2\pi j} \int_{\partial\Omega} \frac{f'(s)}{f(s)} \, ds = \left[\sum_i t_{z_i} + \sum_i t^b_{z_i}\right] - \left[\sum_i t_{p_i} + \sum_i t^b_{p_i}\right]$$

$$= {}^{\#}Z - {}^{\#}P \qquad\qquad (A5.4.1)$$

where $^{\#}Z$ and $^{\#}P$ are respectively the total number of poles

and zeros of $f(s)$ in Ω enclosed by the boundary $\partial\Omega$ traversed

in the positive sense with respect to Ω, and where Ω lies

in the Riemann surface of the algebraic function $f(s)$. Note

that multiple poles and zeros must be counted as many times

as their orders indicate.

The left-hand side of equation (4.1) is equivalent to

the net sum of the clockwise encirclements of a curve set Γ

about the origin of the f-plane where Γ is the image of $\partial\Omega$ under

$f(s)$. Denoting this net sum of clockwise encirclements by

$N(\Gamma,0)$, we finally obtain the required extended Principle of

the Argument in the form

$$N(\Gamma,0) = {}^{\#}Z - {}^{\#}P \qquad\qquad (A5.4.2)$$

Appendix 6.   Multivariable pivots from the characteristic equation $\Delta(g,s)=0$

Consider the characteristic equation

$$\Delta(g,s) = 0 \qquad (A6.1)$$

and assume we have found an approximation for a branch of the characteristic frequency loci about $s=\infty$, as $k\to\infty$, of the form

$$s \overset{\sim}{=} bk^{\alpha} \qquad (A6.2)$$

If ρ is the pivot corresponding to this asymptote then for $k=\infty$

$$s = \rho + bk^{\alpha} \qquad (A6.3)$$

or

$$k = \left(\frac{s-\rho}{b}\right)^{\frac{1}{\alpha}} \qquad (A6.4)$$

The dependence of the closed-loop poles on k is given by equation (6.1.5), i.e.

$$\Gamma(k,s) = 0 \qquad (A6.5)$$

Therefore substituting for k or s in equation (A6.5) we obtain a relationship between s and ρ or a relationship between k and ρ.   Since the order of s will in general be much greater than the order of k in equation (A6.5), we will consider the case of substituting for k.   Substituting for k from equation (A6.4) into equation (A6.5) we obtain the equation

$$Z(\rho,s) = 0 \qquad (A6.6)$$

We require ρ for $s=\infty$, and so we put $s=z^{-1}$ in equation (A6.6) to give

$$Z(\rho,s) = Z(\rho,z^{-1}) = z^{-t}\Omega(\rho,z) = 0 \qquad (A6.8)$$

where t is the maximum order of s in (A6.6).   The equation

$$\Omega(\rho,0) = 0 \qquad (A6.8)$$

then gives the multivariable pivot, ρ.

<u>Example</u>

Consider the open-loop gain matrix

$$G(s) = \frac{1}{s^4-s^3+2s^2-25s+29}\begin{bmatrix} -s^3-11s^2-29s+92 & -20s^2+35s+70 \\ \\ 41s^2-s-91 & 33s^2-170s+118 \end{bmatrix}$$

which has a characteristic equation

$$\Delta(g,s)=(s^4-s^3+2s^2-25s+29)g^2-(-s^3+22s^2-199s+210)g+(-33s+594)=0$$

Using the method described in section 6.2 we find that as $k\to\infty$ the infinite branches of the characteristic frequency loci are given by the following approximations

$$s\backsimeq k \text{ and } s\backsimeq\pm\sqrt{(-33k)}$$

To find the pivot corresponding to the second-order Butterworth pattern $j\sqrt{(33k)}$ we will put

$$s=\rho+\sqrt{(-33k)}$$

giving

$$k=\frac{(s-\rho)^2}{-33}$$

From the characteristic equation $\Delta(g,s)=0$ we obtain

$$\Gamma(s,k)=(s^4-s^3+2s^2-25s+29)+(-s^3+22s^2-199s+210)k+(-33s+594)k^2=0$$

and substituting for k in this we obtain

$$Z(\rho,s)=33^2(s^4-s^3+2s^2-25s+29)-33(-s^3+22s^2-199s+210)(-2\rho s+\rho^2)$$
$$-33s^2(22s^2-199s+210)+(-33s+594)(-4s^3\rho+6s^2\rho^2-4s\rho^3+\rho^4)$$
$$+594s^4=0$$

Putting $s=z^{-1}$ we find

$$\Omega(\rho,z)=33^2(1-z+2z^2-25z^3+29z^4)$$
$$-33(-1+22z-199z^2+210z^3)(-2\rho+\rho^2z)-33(22-199z+210z^2)$$
$$+(-33+594z)(-4\rho+6z\rho^2-4z^2\rho^3+z^3\rho^4)+594=0$$

Therefore

$$\Omega(\rho,0)=33^2-66\rho-33\times22+33\times4\rho+594=0$$

from which we find

$$\underline{\rho=-14.5}$$

Appendix 7. Association between branch points and stationary points on the gain and frequency surfaces

Let us suppose that from an open-loop gain matrix $G(s)$, we have obtained an algebraic equation in terms of the complex gain variable g, and the complex frequency variable s, given by

$$F \overset{\Delta}{=} f(g,s) = 0 \qquad\qquad (A7.1)$$

Let us also consider the following equations involving the derivatives of F:

$$\left(\frac{\partial F}{\partial g}\right)_s = 0 \qquad\qquad (A7.2)$$

and

$$\left(\frac{\partial F}{\partial s}\right)_g = 0 \qquad\qquad (A7.3)$$

The values of s simulataneously satisfying equation (A7.1 and .2) are the branch points on the frequency surface; the values of g simultaneously satisfying equations (A7.1 and .3) being the branch points on the gain surface.

If we consider g as a function of s, the total derivative of F with respect to s is

$$\frac{dF}{ds} = \left(\frac{\partial F}{\partial s}\right)_g + \left(\frac{\partial F}{\partial g}\right)_s \frac{dg}{ds} \qquad\qquad (A7.4)$$

Alternatively we can consider s as a function of g, and obtain

$$\frac{dF}{dg} = \left(\frac{\partial F}{\partial g}\right)_s + \left(\frac{\partial F}{\partial s}\right)_g \frac{ds}{dg} \qquad\qquad (A7.5)$$

Now from equation (A7.1), F is identically zero and therefore its total derivatives must also be zero, so that from equations (A7.4 and .5) we have the following:

$$\left(\frac{\partial F}{\partial s}\right)_g = -\left(\frac{\partial F}{\partial g}\right)_s \frac{dg}{ds} \tag{A7.6}$$

and

$$\left(\frac{\partial F}{\partial g}\right)_s = -\left(\frac{\partial F}{\partial s}\right)_g \frac{ds}{dg} \tag{A7.7}$$

If we have a branch point on the gain surface equation (A7.3) is satisfied and hence from equation (A7.6) we have that

$$\left(\frac{\partial F}{\partial g}\right)_s = 0 \qquad \text{or} \qquad \frac{dg}{ds} = 0$$

which imply that on the frequency surface we have either a branch point or a stationary point corresponding to the gain surface branch point.

Alternatively if we have a branch point on the frequency surface equation (A7.2) is satisfied and hence from equation (A7.7) we have that

$$\left(\frac{\partial F}{\partial s}\right)_g = 0 \quad \text{or} \quad \frac{ds}{dg} = 0$$

which imply that on the gain surface we have either a branch point or a stationary point corresponding to the frequency surface branch point.

References

[1] B.A. Fuchs, and V.I. Levin, "Functions of a Complex Variable", International Series of Monographs in Pure and Applied Mathematics, Pergamon Press, 1961 (translation of 1951 Russian original).

[2] F.M. Ayres, "Theory and Problems of Matrices", Schaum, New York, 1962.

[3]   S.Barnett, "Matrices in Control Theory", Van Nostrand-
        Reinhold, London, 1971.

[4]   G. Sansone, and J. Gerretsen, "Lectures on the Theory of
        Functions of a Complex Variable", Vol. 2, Wolters-
        Nordhoff, Groningen, 1969.

[5]   M.A.Evgrafov, "Analytic Functions", Dover, New York, 1978.

[6]   G.A. Bliss, "Algebraic Functions", Dover, New York, 1966
        (reprint of 1933 original).

[7]   G. Springer, "Introduction to Riemann Surfaces" Addison-
        Wesley, Reading, Mass., 1957.

[8]   K. Knopp, "Theory of Functions", Part 2, Dover, New York,
        1947.

Bibliography

AYRES, F.M., "Theory and Problems of Matrices", Schaum,
New York, 1962.

BARMAN, J.F., and KATZENELSON, J., Memorandum ERL-383,
Electronics Research Laboratory, College of Engineering,
Univ. of California, Berkeley, 1973.

BARMAN, J.F., and KATZENELSON, J., "A generalized Nyquist-
type stability criterion for multivariable feedback systems",
Int. J. Control, 20, 593-622, 1974.

BARNETT, S., "Matrices in Control Theory", Van Nostrand-
Reinhold, London, 1971.

BLISS, G.A., "Algebraic Functions", Dover, New York, 1966,
(reprint of 1933 original).

BODE, H.W., "Network analysis and feedback amplifier design",
Van Nostrand, Princeton, N.J., 1945.

BOHN, E.V. "Design and synthesis methods for a class of
multivariable feedback control systems based on single
variable methods", Trans.AIEE, 81, Part 2, 109-115, 1962.

BOHN, E.V. and KASVAND, T, " Use of matrix transformations
and system eigenvalues in the design of linear multivariable
control systems", Proc.IEE, 110, 989-997, 1963.

BOKSENBOM A.S. and HOOD, R, "General algebraic method applied
to control analysis of complex engine types", National Advisory
Committee for Aeronautics, Report NCA-TR-980, Washington D.C.,
1949.

BRACEWELL, R., "The Fourier Transform and Its Applications",
McGraw-Hill, New York, 1965.

CANNON, R.H., Jr., "Dynamics of Physical Systems", McGraw-Hill, New York, 1967.

COHN, P.M., "Algebra", Vol. 1, Wiley, London, 1974.

DESOER, C.A., and CHAN, W.S., "The Feedback Interconnection of Lumped Linear Time-invariant Systems", J. Franklin Inst., 300, 335-351, 1975.

ELGERD, O.I., "Control System Theory", McGraw-Hill, New York, 1967.

EVANS, W.R., "Graphical Analysis of Control Systems", Trans. AIEE, 67, 547-551, 1948.

EVANS, W.R., "Control System Synthesis by Root Locus Method", Trans. AIEE, 69, 1-4, 1950.

EVANS, W.R., "Control System Dynamics", McGraw-Hill, New York, 1954.

FREEMAN, H., "A synthesis method for multipole control systems", Trans. AIEE, 76, 28-31, 1957.

FREEMAN, H., "Stability and physical realizability considerations in the synthesis of multipole control systems", Trans.AIEE, Part 2, 77, 1-15, 1958.

FUCHS, B.A., and LEVIN, V.I., "Functions of a Complex Variable", International Series of Monographs in Pure and Applied Mathematics, Pergamon Press, 1961 (translation of 1951 Russian original).

GANTMACHER, F.R., "Theory of Matrices", Vol. 1, Chelsea, New York, 1959.

GOLOMB, M.,and USDIN, E., "A theory of multidimensional servo systems", J. Franklin Inst., 253(1), 28-57, 1952.

HILLE, E., "Analytic Function Theory", Vol. 1, Ginn and Co., U.S.A., 1959.

HILLE, E., "Analytic Function Theory", Vol. 2, Ginn and Co., U.S.A. 1962.

JURY, E.I., "Inners and Stability of Dynamical Systems", Wiley, New York, 1974.

KALMAN, R.E., "When is a linear control system optimal?", Trans.ASME J.Basic Eng., Series D., 86, 51-60, 1964.

KAVANAGH, R.J., "Noninteraction in linear multivariable systems", Trans. AIEE, 76, 95-100, 1957.

KAVANAGH, R.J., "The application of matrix methods to multi-variable control systems", J.Franklin Inst., 262, 349-367, 1957.

KAVANAGH, R.J., "Multivariable control system synthesis", Trans. AIEE, Part 2, 77, 425-429, 1958.

KONTAKOS, T., Ph.D. Thesis, University of Manchester, 1973.

KNOPP, K., "Theory of Functions", Part 2, Dover, New York, 1947.

KOUVARITAKIS, B., and SHAKED, U., "Asymptotic behaviour of root-loci of multivariable systems", Int. J. Control, 23, 297-340, 1977.

KWAKERNAAK, H., "Asymptotic Root Loci of Multivariable Linear Optimal Regulators", IEEE Trans. Automatic Control, 21, 378-382, 1976.

KWAKERNAAK, H., and SIVAN,R., "Linear Optimal Control Systems", Wiley, New York, 1972.

MACFARLANE, A.G.J., "Dual system methods in dynamical analysis Pt. 2 - Optimal regulators and optimal servo-mechanisms", Proc. IEE, 116, 1458-1462, 1969.

MACFARLANE, A.G.J., "Dynamical System Models", Harrap, London, 1970.

MACFARLANE, A.G.J., "Return-difference and return-ratio matrices and their use in analysis and design of multivariable feedback control systems", Proc. IEE, 117, 2037-2049, 1970.

MACFARLANE A.G.J., and BELLETRUTTI, J.J., "The Characteristic Locus Design Method", Automatica, 9, 575-588, 1973.

MACFARLANE, A.G.J., and KARCANIAS, N., "Poles and zeros of linear multivariable systems: a survey of the algebraic, geometric and complex variable theory", Int. J. Control, 24, 33-74, 1976.

MACFARLANE, A.G.J. and KOUVARITAKIS, B., "A design technique for linear multivariable feedback systems", Int. J.Control, 25, 837-874, 1977.

MACFARLANE, A.G.J., KOUVARITAKIS, B., and EDMUNDS, J.M.,"Complex variable methods for multivariable feedback systems analysis and design", Alternatives for Linear Multivariable Control, National Engineering Consortium, Chicago, 189-228, 1977.

MACFARLANE, A.G.J., and POSTLETHWAITE, I., "The generalized Nyquist stability criterion and multivariable root loci", Int. J. Control, 25, 81-127, 1977.

MACFARLANE, A.G.J.,and POSTLETHWAITE, I., "Characteristic frequency functions and characteristic gain functions", Int. J. Control, 26, 265-278, 1977.

MACFARLANE, A.G.J. and POSTLETHWAITE, I., "Extended Principle of the Argument", Int. J. Control, 27, 49-55, 1978.

MAYNE, D.Q., "The Design of Linear Multivariable Systems", Automatica, 9, 201-207, 1973.

MEES, A.I., and RAPP, P.E., "Stability criteria for multiple-loop nonlinear feedback systems", Proc. IFAC Fourth Multivariable Technological Systems Symposium, Fredericton, Canada, 1977.

NYQUIST, H., "The Regeneration Theory", Bell System Tech. J., 11, 126-147, 1932.

OWENS, D.H., "A note on series expansions for multivariable root-loci", Int. J. Control, 26, 549-557, 1977.

POSTLETHWAITE, I., "The asymptotic behaviour, the angles of departure, and the angles of approach of the characteristic frequency loci", Int. J. Control, 25, 677-695, 1977.

POSTLETHWAITE, I., "A generalized inverse Nyquist stability criterion", Int., J. Control, 26, 325-340, 1977.

POSTLETHWAITE, I., "A note on the characteristic frequency loci of multivariable linear optimal regulators", IEEE Trans. Automatic Control, 23, 757-76o, 1978.

RAYMOND, F.H., "Introduction a l'étude des asservissements multiples simultanes", Bull. Soc. Fran. des Mecaniciens, 7, 18-25, 1953.

ROSENBROCK, H.H., "On the design of linear multivariable control systems", Proc. Third IFAC Congress London, 1, 1-16, 1966.

ROSENBROCK, H.H., "Design of multivariable control systems using the inverse Nyquist array", Proc. IEE, 116, 1929-1936, 1969.

ROSENBROCK, H.H., "State Space and Multivariable Theory", Nelson, London, 1970.

ROSENBROCK, H.H., "Computer-aided control system design", Academic Press, London, 1974.

SAEKS, R., "On the Encirclement Condition and Its Generalization", IEEE Trans. on Circuits and Systems, 22, 780-785, 1975.

SANSONE, G., and GERRETSEN J., "Lectures on the Theory of Functions of a Complex Variable", Vol. 2, Wolters-Nordhoff, Groningen, 1969.

SHAKED, U., "The angles of departure and approach of the root-loci in linear multivariable systems", Int. J. Control, 23, 445-457, 1976.

SPRINGER, G., "Introduction to Riemann Surfaces", Addison-Wesley, Reading, Mass., 1957.

WHITELEY, A.L., "Fundamental Principles of Automatic Regulators and Servo Mechanisms", J. IEE, 94, Part IIA, 5-22, 1947.

WILLEMS, J.L., "Stability Theory of Dynamical Systems", Nelson, London, 1970.

INDEX

175

Lecture Notes in Economics and Mathematical Systems

For information about Vols. 1–104 please contact your bookseller or Springer-Verlag